Monsieur

Porto betaglet und 21. Bysexme!

Monsieur Linnæus
Professeur de la Bota-
nique
                    à
paid 1ʳ   Upsal
         Stockholm

Letters to Linnaeus

Edited by
Sandra Knapp and Quentin Wheeler

The Linnean Society of London 2009

First published by The Linnean Society of London in 2009

The Linnean Society of London identifies Donat Agosti, John Alcock, Robert S. Anderson, Stephen Blackmore, Joel Cracraft, Peter Crane, David Cutler, Hugh Downs, Joseph Dubois, Malte Ebach, James L. Edwards, Richard Fortey, Nico Franz, Charles Godfray, Andrew Hamilton, Stephen Hopper, Christopher J. Humphries, Sally Kitch, Sandra Knapp, Richard P. Korf, Frank Krell, Leonard Krishtalka, Richard Lane, Ellinor Michel, David P. Mindell, Alessandro Minelli, Gerry Moore, Gareth Nelson, Lynne Parenti, Alan Paton, Norm Platnick, Andrew Polaszek, Richard Pyle, Kevin de Queiroz, Peter Raven, Olivier Rieppel, Gaden Robinson, Sara Scharf, Malcolm Scoble, Ole Seberg, Jorge Soberon, Dennis Stevenson, Tod Stuessy, James E. Sublette, F. Christian Thompson, Gunnar Tibell, Charles A. Triplehorn, Antonio G. Valdecasas, Dick Vane-Wright, Quentin Wheeler, David Williams and Edward O. Wilson as the authors of this work. The authors assert their moral rights.

Design and Text © The Linnean Society of London, 2009.
Preface © The Linnean Society of London, 2009. Sandra Knapp and Quentin Wheeler assert their moral rights.

ISBN 978-0-9506207-9-4

British Library Cataloguing-in-Publication data:
A catalogue record for this book is available from the British Library.

All rights reserved. No part of this publication may be reproduced, stored in a retrieval system, or transmitted in any form or by any means, mechanical, photocopying, recording, or otherwise, without the prior permission of the publishers.

The designation of geographical entities in this book, and the presentation of material, do not imply the expression of any opinions whatsoever on the part of the publishers, the editors, authors, or any other participating organisations concerning the legal status of any country, territory, or area, or of its authorities, or concerning the delimitation of its frontiers or boundaries.

Typeset by Servis Filmsetting Ltd, Stockport, Cheshire   Cover design: Heather Oliver
Typeface ITC Stone Serif   *System* InDesign®   Index: Angie Hipkin

Printed and bound in Italy by Graphicom

The Linnean Society of London
Burlington House, Piccadilly
London W1J 0BF UK
www.linnean.org   Charity Reference No. 220509

Contents

Preface — ix

The Correspondents

Donat Agosti — 1
PLAZI, Switzerland

John Alcock — 5
Arizona State University, Tempe, Arizona, USA

Robert S. Anderson — 11
Canadian Museum of Nature, Ottawa, Ontario, Canada

Peter Artedi (Marcelo de Carvalho) — 15
Universidade de São Paulo, São Paulo, Brazil (Carvalho)

Anders Berlin — 21
From London, England in 1773

Stephen Blackmore — 25
Royal Botanic Garden, Edinburgh, Edinburgh, Scotland

Mark Catesby — 29
From London, England in 1745

Anders Celsius (Leonard Krishtalka) — 31
University of Kansas, Lawrence Kansas, USA (Krishtalka)

Joel Cracraft — 35
American Museum of Natural History, New York, New York USA

Peter Crane — 41
University of Chiacago, Chicago, Illinois USA

David Cutler — 45
Linnean Society of London, London, England

Hugh Downs — 49
Paradise Valley, Arizona, USA

Joseph Dubois — 51
Phoenix, Arizona, USA

James L. Edwards — 53
Encyclopedia of Life, Smithsonian Institution, Washington DC, USA

Georg Dionysius Ehret — 57
From Chelsea, London, England in 1736

Richard Fortey — 59
Ascot, Berkshire, England

Nico Franz — 63
University of Puerto Rico, Mayagüez, Puerto Rico

Charles Godfray — 75
University of Oxford, Oxford, England

Johann Wolfgang Goethe (Malte Ebach) — 81
Arizona State University, Tempe, Arizona, USA (Ebach)

Antoine Gouan From Montpellier, France in 1770	85
Andrew Hamilton Arizona State University, Tempe, Arizona, USA	87
Stephen Hopper Royal Botanic Gardens, Kew, Richmond, England	93
Christopher J. Humphries The Natural History Museum, London, England	99
Pehr Kalm From London, England in 1746	103
Sally Kitch Arizona State University, Tempe, Arizona, USA	105
Sandra Knapp The Natural History Museum, London, England	111
Richard P. Korf Cornell University, Ithaca, New York, USA	117
Frank Krell Denver Museum of Nature and Science, Denver, Colorado, USA	119
Richard Lane The Natural History Museum, London, England	123
Francis Masson South Africa, via London, England in 1775	129
Ellinor Michel International Trust for Zoological Nomenclature, London, England	133
David P. Mindell University of Michigan, Ann Arbor, Michigan, USA	139
Alessandro Minelli Università de Padova, Padua, Italy	147
Gerry Moore Brooklyn Botanic Garden, Brooklyn, New York, USA	153
Gareth Nelson University of Melbourne, Melbourne, Victoria, Australia	157
Lynne Parenti National Museum of Natural History, Smithsonian Institution, Washington DC, USA	163
Alan Paton Royal Botanic Gardens, Kew, Richmond, England	169
Norm Platnick American Museum of Natural History, New York, New York, USA	171
Andrew Polaszek The Natural History Museum, London, England	185
Richard Pulteney From Blandford, Dorset, England in 1782	189
Richard Pyle Bishop Museum, Honolulu, Hawaii, USA	191
Kevin de Queiroz National Museum of Natural History, Smithsonian Institution, Washington DC, USA	199

Peter Raven Missouri Botanic Garden, St. Louis, Missouri, USA	205
Olivier Rieppel Field Museum of Natural History, Chicago, Illinois, USA	207
Gaden Robinson The Natural History Museum, London, England	213
Jean Jacques Rousseau *From Paris, France in 1771*	223
Sara Scharf University of Toronto, Toronto, Ontario, Canada	225
Malcolm Scoble The Natural History Museum, London, England	235
Ole Seberg Natural History Museum of Denmark, Copenhagen, Denmark	239
Jorge Soberon University of Kansas, Lawrence, Kansas, USA	245
Daniel Solander *From Cornhill, London, England (undated)*	249
Dennis Stevenson The New York Botanical Garden, Bronx, New York, USA	251
Tod Stuessy University of Vienna, Vienna, Austria	255
James E. Sublette Colorado State University, Pueblo, Colorado, USA	259
F. Christian Thompson Agriculture Research, USDA, Washington DC, USA	263
Gunnar Tibell Uppsala University, Uppsala, Sweden	267
Charles A. Triplehorn The Ohio State University, Columbus, Ohio, USA	273
Antonio G. Valdecasas Museo Nacional Ciencias Naturales, Madrid, Spain	275
Dick Vane-Wright Canterbury, Kent, England	281
Arnout Vosmaer *From Den Haag, The Netherlands in 1760*	287
Quentin Wheeler Arizona State University, Tempe, Arizona, USA	291
David Williams The Natural History Museum, London, England	297
Edward O. Wilson Harvard University, Cambridge, Massachusetts, USA	311
Linnaean works cited by the correspondents	317
Picture credits	318
Index	319

CAROLI LINNÆI

Equitis De Stella Polari,
Archiatri Regii, Med. & Botan. Profess. Upsal.;
Acad. Upsal. Holmens. Petropol. Berol. Imper.
Lond. Monspel. Tolos. Florent. Soc.

SYSTEMA NATURÆ

Per

REGNA TRIA NATURÆ,

Secundum

CLASSES, ORDINES, GENERA, SPECIES,

Cum

CHARACTERIBUS, DIFFERENTIIS, SYNONYMIS, LOCIS.

TOMUS I.

Editio Decima, Reformata.

Cum Privilegio S:æ R:æ M:tis Sveciæ.

HOLMIÆ,
Impensis Direct. LAURENTII SALVII,
1758.

Preface

"With the exception of Shakespeare and Spinoza, I know of no one among the no longer living who has so strongly influenced me"
—Johann Wolfgang Goethe (of Linnaeus)

"Linnaeus was actually a poet who happened to become a naturalist"
—August Strindberg

The idea for this book arose from our shared desire to acknowledge the 250th anniversary of the publication of *Systema Naturae*, the work in which Carolus Linnaeus (or Carl von Linné) first used a uniform naming system for *all* of life. It is difficult to overstate the importance of this event in the evolution of human communication about the natural world, eclipsed only by the technical milestones of moveable type in the 15th century and the Internet in the 20th. The printing of the 10th edition of *Systema Naturae* in 1758 marked the birth of 'modern' classification and nomenclature…at least in most of zoology. An exception is the spiders for which a work by Linnaeus' contemporary Carl Clerck[1] published in 1757 is accepted as the starting point. For plants, binomial naming began with Linnaeus' *Species Plantarum* published in 1753. Even with the restart of bacterial naming in 1980, there is no denying the nonpareil influence that Linnaeus had and continues to have on biological nomenclature and classification. Because some contemporary authors see the Linnaean classification system as anachronistic and of diminishing relevance, we believe it is timely to take stock of the impact of Linnaeus' innovations and to reflect carefully on the potential consequences of moving away from a system that has proven itself for two and a half centuries. Perhaps a day will come when it is a good idea and a right time to abandon the Linnaean system, although we doubt it. It is simply too useful, too simple, too adaptable, and too well conceived to cast aside. It is simply genius. This fact and not some collective sentimentality in the taxonomic community is the reason for an enduring use of a Linnaean system.

Excellent biographies of Linnaeus have been published[2]. Histories of taxonomy and nomenclature have been written and the international decisions concerning nomenclature continue to expand and improve upon formal rules for the coining and use of scientific names. Although a definitive history of taxonomy, particularly one including the tumultuous 20th century, has yet to be written as a fitting tribute to Linnaeus that is not our mission here. Ours is more humble, a little less scholarly, but a little more humane and idiosyncratic.

We invited an eclectic set of authors, some taxonomists, some not, to reflect during 2008 on what Linnaeus has meant to them personally and professionally. In so doing we knew that we would elicit a wide variety of insights into how the Linnaean influence has fostered good science and the development of good scientists. We hoped in addition to reveal something of the human side of science rarely seen in professional publications or essays. The authors did not disappoint and the letters range from sober to hilarious, from mini-essays on contentious topics in the field to flights of fancy, from long and detailed to short and sweet. They tell the story of the fits, starts, successes, failures, frustrations and joy of the science of understanding nature that has gone on in the 250 years since the publication of *Systema Naturae* in 1758. We have only edited the letters very lightly, and of course not edited those sent to Linnaeus in his lifetime at all! Format, spellings and turns of phrase have been left in each of the letters as the authors sent them—this has unexpectedly revealed to us that letter-writing as an art is as alive and well today as it was in Linnaeus' time. Many common themes and concerns emerge, but we are delighted that contributors revealed so much diversity and humanity, and above all good humour.

Something of the joy of doing science is lost when our discoveries are passed through a filter of feigned objectivity designed to separate out the cold facts and leave behind the human motives. We believe there is nothing wrong with being human. In fact we have both tried with general if uneven success to do so ourselves. That science is done by humans introduces mistakes, biases, and the full range of human emotion. Perhaps the most amazing thing about science is that this does not matter, at least not in the long run. Because good science appeals to further observations of nature to refute or corroborate itself, the human factor is not permitted to stand in the way of the path towards discovering the truth of the matter.

Science has the almost miraculous property of rising above, or at least eventually compensating for, human failings. The very best science is almost invariably associated with strong human emotion. At its extremes it may be a narcissistic drive to show off genius or a loathing desire to show that someone else's ideas are wrong. At its best science is done with passion and *con amore*.

E. O. Wilson[3] wrote: "Historians of science often observe that asking the right question is more important than producing the right answer. The right answer to a trivial question is also trivial, but the right question, even when insoluble in exact form, is a guide to major discovery." Linnaeus asked the right questions about the living world. Questions we are still trying to answer; questions that are as relevant and impactful today as they were in 1758.

Linnaeus asked "What species exist?" Our estimate of the answer to the related numerical question "How many species are there?" has changed dramatically from the approximately 10,000 species he predicted to ten or more millions of species, only 10% of which are currently known. Answering the 'what' question is by no means a done deal. Taxonomists have discovered, described and named about 1.8 million species and another 20,000 or so are added each year.

The big questions in taxonomy[4] can be traced back to Linnaeus if only by implication: What are species? What species exist? How are they related? Where are they distributed? What is the history of the origin and diversification of their characters? How are organisms best classified? These are among the most basic and essential questions for understanding biodiversity yet they remain woefully incompletely answered.

A lot has changed since the mid-18th century yet the Linnaean system continues to serve biology well. Although international rules for botanical and zoological nomenclature have changed and expanded they remain founded on the fundamental concepts given by Linnaeus. Binomials have proven up to the task of naming millions of species using informative and memorable words. The simple, adaptable genius in Linnaeus' binomial system easily explains its uniform adoption and continued use. What is more surprising are the ways in which the Linnaean system seems pre-adapted to meet contemporary needs and opportunities. For example, Linnaeus' categorical ranks are perfectly suited to reflect the nested sets of relationships revealed in a cladogram. His typographic practice of capitalising the genus name and not the specific epithet (as in *Homo sapiens*) augmented by our modern convention of putting the name in italicised print allows clever algorithms to crawl through hundreds of pages of scanned publications determining what scientific names appear and how many times.

Linnaeus gave to us a beautiful, expressive, informative language with which to share with one another our observations and knowledge of the world's species. A language so simple and yet so powerful that it has proven equally effective at expressing incredibly divergent world views. Creationists, quinarians, evolutionists, neo-Darwinians, pheneticists, cladists, and molecular phylogeneticists have all relied on Linnaean names to express their ideas with equal clarity. What more could one ask of a simple system of words?

The world is a wondrous and beautiful place in large measure because of the diversity of life on our planet. Linnaeus gave voice to biodiversity and our exploration and admiration of it. Without this vocabulary we would be at a loss to explain our knowledge of nature with any degree of accuracy. In the light of the accelerating loss of biodiversity, rapidly changing environments worldwide, and increasing frequency of introductions of non-native species, it is imperative that we accelerate our attempts to answer Linnaeus' questions. For our own sake and that of generations to follow we must establish a credible knowledge of what remains of life on Earth. Unless we have a good baseline how can we possibly detect or monitor changes in biodiversity? Because species and higher taxa are hypotheses that are continually improved through time it is equally important to document earth's species as a record of taxonomic and phylogenetic diversity. Taxonomists for generations to come will refer to collections and associated information to continue this process of improving our concepts and classifications.

xii *Letters to Linnaeus*

Linnaeus has touched everyone who has been curious enough about nature to learn a scientific name. Hikers admiring spring wildflowers, gardeners, fishermen, hunters, conservationists, explorers, and of course field biologists of every stripe. Many of us have used Linnaean names since childhood, what five-year-old does not know exactly what is meant by *Tyrannosaurus rex*? The rote memory of a few Latinised dinosaur or tree or bug names can welcome any child or amateur or aspiring student into the rarified world of science. Linnaeus has touched every biologist in a meaningful way, even if they have never taken pause to recognise it. Linnaeus has given everyone who has ever been moved deeply by the beauty of nature a voice with which to express that awe and admiration of the diversity of life.

We thank the contributors who have taken time to reflect upon the influence and impact of Linnaeus on their lives and science and to express those thoughts in the form of letters written to Linnaeus. It is hard to write to someone you don't really know personally, especially across the centuries. Shannon Keen (ASU) and Leonie Berwick (Linnean Society) organised us to make it all happen, without them we would still be thinking what a good idea this book might be. At the Linnean Society of London, Ruth Temple read the manuscript and Ben Sherwood and Lynda Brooks helped with Linnaean history, Charlie Jarvis of the Natural History Museum provided much information about Linnaeus and his times and Pia Wilson deciphered the handwriting and translated the Swedish in the 18th-century letters to Linnaeus. We also thank the Linnean Society who allowed us to pursue our vision of a book at the crossroads of biology and the humanities as a tribute to Linnaeus and commemoration of the world-altering publication of *Systema Naturae*.

Sandra Knapp

Quentin D. Wheeler

Rue Linné, Paris, August 2008

Notes:

[1] Clerck, C. 1757. *Svenska spindlar, uti sina hufvud-slågter indelte samt under några och sextio särskildte arter beskrefne och med illuminerade figurer uplyste*. Stockholm.

[2] Blunt, W. 2002. *The Compleat Naturalist: a life of Linnaeus*. Frances Lincoln Ltd., London.

[3] Wilson, E. O. 1998. *Consilience: the unity of knowledge*. Knopf, New York.

[4] Cracraft, J. 2002. The seven great questions of systematic biology: an essential foundation for conservation and the sustainable use of biodiversity. *Annals of the Missouri Botanical Garden* 89:127–144.

Dear Carolus

I guess we share something, the need for change in a dismal situation. Whilst I understand that you were looking for new food sources that could stem famines and feed a growing population, we face problems of a truly global nature about which we are only beginning to learn: climate change, competing priorities for land use, loss of biodiversity, increase in population, and not least of all the illusion that by looking at life in the most remote corner of the world that we know it. Google might be the epitome of our times: anything that can be assembled using machines is available, giving us an unprecedented chance to look at trees in a forest in which no one from our world has ever set foot. But we hardly know what it means in a wider context.

Ants comprise much of the biomass of tropical forests. Leafcutter ants (the genus *Atta*) have many castes.

Anything that can not be accessed by machine is not part of this new world. In your time, you could assume that you were able to synthesize the entire knowledge in your field. I assume that if you had had computers and the internet in your day, you would have wanted to make your *Systema Naturae* editions immediately accessible. After all, that's why you published them.

Today, we talk of about 6,000 active taxonomists, well over 100 million pages of printed legacy publications, and a race to produce the ultimate '*Systema Naturae*, EDITIO MMC', marketed with more palatable names such as 'All Species', 'Encyclopedia of Life' or 'Catalogue of Life'. This all can not be digested by one or few people, but we do have machines that could assist in this task. The problem is that most of our knowledge is not accessible by machine, but needs to be converted into machine readable format and adequate resources are not available, including taxonomists' expertise. But even if this might happen, it will not, because of copyright to which our community is consenting tacitly.

There is another problem. Our community at the moment has a huge drive to administer and control knowledge on species: to create the authoritative system. Unlike other mega-science programs, we are not launching a huge global research program for the purpose of documenting the nature of our living world's organisms. Such large projects yield tangential benefits. Originally, NASA and other space agencies spent huge amounts of research money to build systems to observe our world from space in order to understand global land cover and land use change. Such technology and insights were the basis for Google Earth. This remarkable tool is a byproduct of data being collected for a science project, and databases and communication tools that were set up to allow decentralized analyses.

Since we first talked about biodiversity and the biodiversity crisis during a conference in Washington in 1986, we have lost 20 years of research and innovation opportunity. We still operate like we did 20 years ago and, in fact, as much we have over most of the last century. We might have the Internet, but almost anything really innovative is based on personal initiatives, and our community didn't get together to create a research curriculum in the same way that the NASA Land Cover Community did. In that respect, having DNA sequences is just another character set, and so called e-publications for rapid dissemination do not change the *modus operandi* of the established paradigm. We cannot get the publications because of copyright, we cannot read them by machines which are needed to process millions of pages, we cannot know what the underlying observations were because we do not make use of possible linkages to various databases.

In having the NCBI-maintained GenBank and the Global Biodiversity Information Facility (GBIF) we find hope. If we could base all our future publications on a few principles, we might get a step closer to a new synthesis. That would require that all content is open access (we mandate deposition of all our forthcoming taxonomic publications in open repositories), all names are registered in respective name servers like ZooBank; all the material used is documented with links to respective specimen databases using globally unique identifiers; all the illustrations are open access and linked; all characters including gene sequences re-linked to specimen data; all the specimen data includes high resolution geographic data derived from global positioning systems.

However, there are two additional issues which make me wonder: who is really using our new synthesis and why is the current focus so much on data management and not on a proper research program needed to describe the world's species? Conservation is hardly using our data so far, and from now on will spend most of its efforts on mitigating the effect of climate change. Needless to say, they hardly share their data.

Our community is not very imaginative. It too easily settles for the current situation. Copyright exists. We just accept it and miss out on a huge amount of decisive information. Why not change our publishing models to make it completely an online business, linked to all the respective databases from the start? In many fields, imagination drives innovation. Yet we accept that software is prohibitively expensive, that the community cannot or will not produce standardized images. We are not even able to build the most basic name server for zoology that would produce at last a list of available names *sensu* our Code, because we cannot, or do not want to, provide funding for it. Part of the problem is that our science is not competitive enough. Our funders have not been convinced of the importance of such a basic task. Those most likely in a position to find such funding—a complicated and tedious process over many years—are more happy to imitate Wikipedia rather than imagine, plan, and run successful and innovative large-scale projects.

I am disillusioned. You had an idea that was simple and eventually succeeded because it proved a very efficient naming and reference system. Combined with the Web's possibilities it could radically change our understanding of the world's species. Somehow this seems not to be happening. Perhaps one ought to allow for more time. Fifteen years of Web evolution may not be enough.

Maybe one ought to be optimistic, trusting that the Web will allow unexpected developments. Knowing how many species are on planet Earth is still one of the big unsolved questions in science and might yet come into fashion. Maybe one should take joy from developing and implementing visions on how such a new system ought to work, and make it work, even if it takes on only a minute fraction of the challenge ahead—one you took on 250 years ago.

All the best

Donat

The case of *Pheidole* ants (here *Pheidole longispinosa* Forel) is typical for the bias even of leading scientists towards loyalty to their publishers rather than free access to and widest possible distribution of scientific data through Open Access, not to speak of new forms of publishing (Nature 424:985, 2003). This issue undermines all the efforts to put taxonomy back on the scientific and conservation agenda.

Dear Carl Linnaeus,

Although I am not a taxonomist, I have many reasons to celebrate your brilliant idea that each and every species should have its own unique scientific name available to all scientists the world around. Let me supply one example taken from my research, which has to do with the mating behavior of a wide assortment of insects, especially those that fly to hill-tops to seek out mates. For many years now I have visited a very modest mountain close to my home here in Phoenix, Arizona, where insects ranging from hefty tarantula hawk wasps (*Hemipepsis ustulata*) to minute ants, such as *Pogonomyrmex pima*, gather to find a partner and engage in sexual reproduction. Although all these hill-topping species share the convention of using a prominent topographic feature as a mating rendezvous site, the males of the various species use a diversity of tactics to achieve their goals. Some are highly territorial, for example, while others roam about the peak without exhibiting any aggression toward their fellow males.

Explaining the adaptive significance of hill-topping in all its diversity is the major goal of my research. To achieve this goal requires that I know exactly what species I am observing. For one thing, if I did not have a scientific name to insert in the title of my paper, I probably would not be able to publish my findings, and rightly so. Without the correct name, I could not be certain whether the species in question had been previously studied and so would be unable to compare my findings with any that had already been reported in the scientific literature. Without the right name, I would also not know for sure what the close relatives of this species were and so could not compare the mating system(s) of these species with the insect that I had chosen to investigate. Comparisons among members of the same genus are often extremely useful in identifying the key ecological factors that were responsible for the behavioral differences among closely related (genetically similar) species. Without the right name, therefore, my paper would properly languish in a file cabinet, where it could do little to advance my career or to help us understand the evolution of insect mating systems.

But there is another reason why getting the right name on one's subjects is so important for a behavioral biologist like me, which is that how we interpret what we see can be entirely dependent on whether the population we have been watching is composed of one species as opposed to more than one. This point was brought home to me forcefully during the late summer of 2006 when I was regularly visiting my favorite mountain-top, Usery Peak. During the suffocatingly hot month of August, field work in Arizona offers a real challenge to the human thermoregulatory system. Each time I crawled slowly uphill to the top of the mountain, which fortunately is more hill than mountain, I had reason to doubt the wisdom of my research endeavors. On the other hand, having recently discovered males of a small but attractive green-eyed wasp on some of the little trees and shrubs growing in the highest parts of Usery Mountain, I wanted to learn what they were up to. I quickly learned that as a general rule only one specimen occupied any given plant. When another wasp appeared, the male left his perch to meet the intruder, resulting in a brief but rapid chase, and concluding when one male flew back into the plant to perch on a thin limb of a creosote bush or palo verde tree. True, a few trees had two or maybe even three resident males but I supposed that the coexistence of these individuals stemmed from the relatively large size of landmark sites they had chosen to defend.

I collected a couple of specimens of the wasps for later identification by an expert, although initially I did not know who this expert might be, and I carried on with my observations. My key research tactic was to mark individuals that I caught in the set of defended trees and shrubs. When I released these males, each with its own distinctive dots of paint pen colors on its thorax, they often came back to the place where they had been before I had interrupted their activities. In this way, I found that most males were site faithful individuals, which returned over several days to one shrub or tree, offering evidence that these males were defending prominent hill-top perch sites, presumably because receptive females sometimes arrived at these perch plants to be mated. This pattern is one shared by many other hill-topping species.

Things were going well with the gradual but steady accumulation of data on resightings and male-male interactions when on one broiling August afternoon I noticed that one of the wasps in my net had a red and black abdomen whereas I seemed to remember that the wasps I had been catching and marking to that date had entirely black abdomens. Whoa! A subsequent check of my already marked wasps revealed that some had black abdomens but others had black and red abdomens. Because I had not differentiated between the two types as I was collecting data on their behavior, I might well have been putting the results for two different species in the same basket, which did not seem wise. On the other hand, it was possible that perhaps the two types were merely two different color forms of the same polymorphic species. Variation of this sort is commonplace in the natural world. But in either case, it was clear that I had to take action quickly to get a definitive verdict on the wasps' identities.

In these days, Carl (if I may be familiar), skilled taxonomists are disappearing faster than the dinosaurs because as older scientists retire they are not replaced by younger specialists. As a result, very few workers remain that can help someone like me when I need a name appended to a wasp specimen. But after inquiries, a colleague suggested that I contact Wojceich Pulawski at the California Academy of Sciences and, when I did, he kindly agreed to take a look at the material I would send him. Pulawksi is an expert on what used to be called sphecid wasps, but are now placed in the family Crabronidae. I knew from past experience that I was studying "sphecids" but I sure didn't have a clue about which of the 1,400 North American species of this family I had captured.

Pulawski worked quickly. He told me that I had sent him, not one, not two, but three species, all members of the genus *Tachytes*. The black bodied males were *Tachytes ermineus*, but the partly red bodied males included representatives from two species that Pulawski told me are so similar in appearance that they could only be distinguished by examining their genitalia under the microscope. Frankly I was dismayed at this information because it is far more convenient to work with species that can be visually identified in the wild. But at least Pulawski's work told me that I had to pay close attention to the color pattern of the wasps I was marking. Furthermore, I had to see if there were some consistent behavioral differences that might enable me to distinguish between the wasps with red abdomens without dissecting them.

A marked male of *Tachytes ermineus* Banks (Crabronidae) from Usery Peak

In the following days, I came to see that there did indeed appear to be two kinds of males with partly red abdomens. Some perched several meters up in certain palo verdes and creosote bushes, where they occasionally interacted briefly with the black *ermineus*, which also liked to sit in the center of trees and shrubs just below their crowns. Males of the putative third species perched on the ground or near the ground on the outermost twigs and leaves of jojoba shrubs. The low perching males were also noticeably smaller than their cousins in the tops of trees. I collected an additional set of red-abdomen specimens and shipped them off to Pulawski, feeling confident that he would give different names to the males taken from these two very different locations.

He did not.

Instead, according to Pulawski, all eight wasps in the second batch that I sent him were members of a single species, a surprising and unwelcome conclusion as far as I was concerned since I had managed to convince myself that there were two species, each with its own perch preferences. In the face of Pulawski's report, I wondered if the slightly smaller, lower perching males actually belonged to a previously unrecognized species nearly identical in external form to the palo verde perchers. A more likely possibility was that the larger males of the red-abdomen species monopolized the prime territorial spots high in desert trees and shrubs, leaving the smaller males of this putative species to do the best they could at inferior sites close to the ground where they at least would not be assaulted or chased off by larger, more powerful opponents. In other words, how I interpreted my observations was hugely dependent on the name(s) assigned to the wasps I had been catching, marking and watching.

Because of my uncertainty about what was going on, I collected a new batch of wasps for shipment to Dr. Pulawski and he continued to patiently work his way through the specimens. As he did so, he himself became confused (he later told me), so much so that he discarded the formal identification key to *Tachytes* that he had been using and went back to the original published description of one of the species that he thought I had sent him. It was then that he realized that the key he had been using was leading him astray and making life much more complicated for him (and for me) than it had to be. Eventually Pulawski was able to tell me that there were indeed two species of *Tachytes* with red abdomens after all, and my notes told me that these two species corresponded to the two behavioral types, one that perched on or very close to the ground, while the other perched from three to 10 feet high, typically in the center of palo verde trees.

What a relief. I could now eliminate the hypothesis that the wasps with red on the abdomen were a single species with alternative mating tactics (a widespread phenomenon, by the way, among the insects and other animals). Instead I felt confident that I was indeed studying three species whose males probably differed in the height and location of their territorial perches. As the number of days of observation added up, I derived firm evidence that *T. ermineus*, *T. spatulatus* and *T. sculleni* did indeed subdivide the landmark plants they occupied with favored perches ranging from high in the center of palo verdes to low down on the outer branches of creosote bushes (or even the bare ground by a shrub). Moreover, I found that although some plants had up to three males present, no perch site held more than one of any given species. Thus, I concluded that all three species exhibited the standard territorial hill-topping mating system.

Having made progress in describing the natural history of three species of *Tachytes*, I was then able to determine whether my observations were novel, which I did by searching the scientific literature for any mention of the genus. In so doing, I assembled a very modest list of four papers that made mention of male behavior. The entomological researchers who had written these papers described, generally very briefly, how males of *T. amazonus, T. distinctus, T. intermedius*, and *T. tricinctus* appear to locate mates. (To my surprise, I learned that I myself was an author of one of these papers, having penned a few notes on *amazonus*, a species that I had watched many years ago near Buenos Aires, Argentina.) Because there are 394 species of *Tachytes* in all (according to www.calacademy.org/research/entomology/Entomology_Resources/Hymenoptera/sphecidae/Number_of_Species.html), 390 remained fair game for those of us who wanted to discover something entirely new about male behavior in this genus. I was happy that I had *ermineus, spatulatus,* and *sculleni* all to myself.

Usery Peak in south central Arizona

The four *Tachytes* for which at least something about male behavior was already known were species in which males assemble near places where many females had made their underground nests the previous year. In these places at the appropriate season, many females eventually make the transition from pupa to adult, after which they burrow up to the surface and fly away. These fresh females are almost certainly receptive to males that catch them soon after their emergence. Accordingly, males of these species perch either on the ground or close to it, leaving their perches to chase potential mates flying nearby. The fact that I did not see a single nest of any *Tachytes* in the Userys, let alone aggregations of nesting females, coupled with the male preference (in two species) for perches high in the center of trees and shrubs suggests strongly that in the wasps I studied, males were employing the standard territorial hill-topping regime. Indeed, the two species of *Tachytes* wasps that perch high in palo verde trees use exactly the same set of plants favored by other hill-topping insects, including the tarantula hawk wasp and *Pogonomyrmex* ants that come to Usery Peak much earlier in the year. The *Tachytes* males are probably also waiting for virgin females to rendezvous with them at these conspicuous landmarks. Perhaps these males employ their approach because their females do not nest in groups but instead are scattered widely through the desert, making it unprofitable for males to attempt to locate emerging females. If so, and if unmated females have evolved a tendency to fly to topographic landmarks, then male hill-topping behavior would make adaptive sense.

Even though I do not yet know whether my hypothesis is correct, I am pleased at the payoff for climbing Usery Peak during the summer, painful though the uphill haul was at times. Although it is not at all surprising to encounter insect species whose life stories are unknown, there is joy in getting to know even rather ordinary insects as individuals belonging to a particular species, a pleasure in the discovery of something new as well as in the feeling of accomplishment upon the communication of one's findings with other scientists (admittedly the number of my colleagues fascinated with *Tachytes* wasps is small but there are at least a few such biologists). All of these outcomes are entirely dependent upon the Linnaean system, for which I am extremely grateful to you, Carl Linnaeus, and to your modern academic descendants, including Wojciech Pulawski. Thanks to you all.

John Alcock

March 1, 2008
Dear Professor Linnaeus:

It has been a short 250 years since the publication of *Systema Naturae* and no doubt you would be asking yourself, were you here today, what has been accomplished in terms of knowing the diversity of life on Earth over this time. Well, the answer would be akin to a double-edged sword. Certainly we have accomplished a lot, with the formal description of more than a million different species, but increasingly we are coming to realize that this may only be an insignificant number as far as compiling a list of species is concerned. When you proposed your system of binominal nomenclature and included the first species in the genera that you recognized, you gave them names that emphasized their distinctive characteristics and made for their easy recognition. However, as we have explored more remote and inaccessible areas of Earth, have used novel methods of collection, and have revealed a world of cryptic species through the use of behavioral, genetic and molecular methods in systematics, we have discovered a wealth of species that have exceeded the number thought possible by yourself and even systematists who were working 100 years ago. In many groups your names no longer enable the recognition of the species as now many species are known to share similar features to those you considered unique at the time. Let me give you some examples from my own work these past 15 years on beetles in the family Curculionidae from the tropical Americas.

Many weevils are tiny, like this specimen of *Protapion trifolii* (L.) from the Linnaean collections

I study weevils that live in the soil and leaf litter and for these past years have made it my personal life quest to, as I jokingly say, "pass all of the Central American forests through a Berlese funnel". Invented by the Italian Antonio Berlese in the late 1800s, this simple piece of equipment uses heat to extract all of the minute arthropods that inhabit soil and leaf litter from the mass of debris in which they live. These creatures are highly specialized for this way of life, are found nowhere else and are never found by other methods. Consider the weevil genus *Theognete*, described at the turn of the 20th century by a wonderful coleopterist named George C. Champion from the British Museum of Natural History in London, England. When Champion described *Theognete* he possessed but two specimens, one from Mexico and one from Guatemala. He placed these in one species which he called *Theognete laevis*. And so it stood for the next almost 100 years, a small genus of but one species. Well, with increased travel to Central America and Mexico and the increased use of Berlese funnels by others as well as by me, the numbers of *Theognete* specimens began to grow. In early 1990 I started a study of the group expecting to add a few species to the solitary one already known. My first discovery was that Champion's two specimens were of different species and in fact just about every specimen I examined was of a different form. Species began to accumulate and today, in 2008, in a manuscript being prepared for publication later this year, I recognize 98 species in the genus. Many of these are limited to single mountain tops. This is not a unique case for the genus *Theognete*. Other weevil genera such as *Acalles, Eurhoptus, Tylodinus, Anchonus*, and *Dioptrophorus*, to name a few, are similarly under-represented in our present day counts of numbers of species. The same is true for nearly all other insects that live in tropical forests, especially those in the leaf litter. In simple words, the diversity of life on Earth is proving to be magnitudes greater than what you ever expected it to be.

The correspondent and a colleague running a roomful of Berlese funnels in Mexico

Since your time, humans have developed the capacity for flight and in fact have taken to the skies and into outer space, beyond the safety of Earth's atmosphere. Humans have actually walked on the moon! Now, the large science agencies that support space exploration are talking about trips to other planets, most likely Mars as it is closest to Earth, but also because it is the most likely to support life. Yes, life! Scientists are willing to spend an incredible amount of money (billions of American dollars) on these fantastic ventures. No doubt, you would find, as I do, that such expenditures are ridiculous when it is becoming increasingly evident that we know so little about life on our own planet! New and fantastic discoveries are being made all the time.

There are now almost 7 billion people in the world and as you might expect, they take up a lot of space and eat a lot of food, and they make more people at an increasingly faster rate. Many forests are almost gone, many large and once plentiful creatures are teetering on the verge of extinction, once productive fisheries are depleted, and more and more of the world's natural spaces are being destroyed to meet the needs of the world's population and we think we may have disrupted the global climate to the extent that we may never be able to get things back to normal. As these forests and other habitats are lost, so are the species of living things that live there. As many of these creatures live nowhere else, they are lost forever. Extinct! Many of them are irretrievably gone, even before we can give them a formal scientific name. I find this unbelievably frustrating, as I believe, so would you.

So whereas we are finding out more and more about the world and its rich diversity of life, we are under increasing pressure to find the resources to be able to carry out our studies at a time which many scientists say is perhaps the greatest period of extinction ever. It's an exciting time to be living but it is also a time to reflect on what the implications are for human life on the future of the planet and the many living things with which we share the world.

Respectfully yours,

Robert S. Anderson
Canadian Museum of Nature
PO Box 3443, Station D
Ottawa, ON. K1P 6P4
Canada

Artedi (Carvalho)

Carl Linnaeus
Uppsala, Sweden

My dearest Carl,

I hope I am not entirely forgotten as I write to you from afar to bring you up to speed on your impact on ichthyology—or should I say *our* impact? Yes, it has been quite a while, and although our interaction was relatively brief, lasting only six years, and somewhat contrite, much benefit world over has resulted from it.

I am sure you remember me well—who can forget that fateful September night when I stumbled and drowned in a murky Amsterdam canal? Perhaps the exhilaration of examining Albertus Seba's collection of fishes, or the elation in anticipation of the completion of my own work on the 'natural organization' of fishes (the *Ichthyologia*, subsequently edited and published by you on my behalf, in five parts, in 1738), left me in a careless, dazed disposition. My work was in its infancy then, as was yours, but nonetheless ichthyology did indeed blossom under your initial guidance, even if you were initially guided by me.

The Butterfly Ray *Gymnura altavela* (L.) (Gymnuridae) was first described by Linnaeus in the 10th edition of *Systema Naturae* (1758)

Your system of classification, undemanding as it may have seemed, has brought much more than a sense of 'orderliness' to the natural world—it has fostered a universal language for what became known as *taxonomy*, itself a foundation of the more encompassing subject of *comparative biology*. From such meager beginnings in 1735, your *Systema Naturae* eventually matured from a 12 page folio-sized pamphlet to achieve almost universal adoption by its much expanded twelfth edition (1766–1768) of well over two thousand pages. As later recognized by another great naturalist, Georges Cuvier, "[i]n simplifying natural history . . . [you] inspired a liking for it generally", and, concerning ichthyology in particular, ". . . the true naturalists writing immediately after [you] . . . were in entire submission". It may be true that binomial nomenclature was perhaps my own intellectual child—or was the idea permeating the academic milieu of our time, like so many others?—but a classificatory structure with a straightforward system of ranks was your greatest methodological insight.

In terms of the natural groups of fishes, dear Carl, may I, with all modesty, continue your education? You relied very heavily upon my own work in the first (1735) edition of your *Systema Naturae*, maintaining my five groups (Malacopterygii, Acanthopterygii, Branchiostegi, Chondropterygii, and Plagiuri) and changing these only in your 10th edition (1758). This demonstrated how much you had learned by then, correctly removing the cetaceans (Plagiuri) from the Pisces and placing them in your Mammalia (Cete); however, your new group, Nantes, strangely allocated to Amphibia next to the Reptiles, seemed to me an odd assortment, with the anglerfish (which I had placed in Branchiostegi) thrown in with the sharks, rays, chimaeras, lampreys, and sturgeons. (Do they really have lungs?) But so much has changed since then, and nowadays some 27,500 species of fishes are known, with many thousands more awaiting discovery and formal binomial description. This serves to gauge the success of your system—in our day, we knew at most only a few hundred varieties of fishes, but that figure has increased by more than two orders of magnitude after 250 years of uninterrupted work, and with your classificatory edifice serving as backdrop throughout.

Botany, dear Carl, had you as its master, and in particular the 'procreation' of plants. One later commentator even described your system of plant classification as a "lyrical dollop of sexual innuendo", for reproduction revealed to you, in your own words, "the secret working-plan of the Creator". But even though fishes never did captivate you in any special way—most of your efforts having consisted of descriptions of private 'cabinets'—many of your students ("apostles") developed a keen interest in them, especially Daniel Solander and Pehr Forsskål. Their travels, observations, and specimens procured have greatly enhanced our appreciation of natural history. I must add here, sadly, that this very positive cascading effect, so essential to build a mounting reference system for the natural sciences, is suffering from depreciation in the 21st century, and at a time when it is most needed. Unexpectedly, this is even playing out in some wealthier nations, where naturalists specializing in this kind of work are having trouble obtaining lasting 'patronage'.

Ironically, some of the major methodological problems that frustrated our early efforts continue to afflict workers even some 275 years later: how do we best interpret variation among specimens when delimiting species, especially in those groups with much individual variation or that display only scanty divergence between species? This is a major obstacle, for example, in sharks and rays: in some freshwater potamotrygonid stingrays of South America (remarkable beasts we had no clue even existed), no two specimens completely match in external features and coloration, not even putative conspecifics of the same size and sex that were sampled together. Dare I say, unraveling the natural organization of life remains, in many ways, a Linnaean conundrum!

Undoubtedly, many challenges persist in the discovery of the true reference system of Nature. Some commonplace difficulties in modern fish taxonomy were mostly unforeseen back in our time: complex metamorphosis in many fish species (bizarre larvae can be very distinct from adults!), morphologically diverse sexes, highly intricate development and morphology, and so forth. To muddle matters even further, many characters occur independently in unrelated groups. This rendered my own novel system of careful description and identification of fish specimens, based, in part, on the position and number of fins and their internal supports, sometimes unreliable. But the classification of fishes has since evolved considerably due to a superb technique in which the skeleton is stained and the overlying flesh is made transparent. Such 'cleared and stained' specimens revealed a maelstrom of dermal bones that eventually provided a more stable path expounding the natural affinities of fishes.

This leads us to another great intellectual pillar in the natural sciences—the conceptual improvements of the so-called 'Hennigian revolution', which took place chiefly from the 1950s onwards. A deep-thinking insect specialist named Willi Hennig advanced a rigorous paradigm to identify reliable characters—those sound evolutionary indicators of affinity—while at the same time teasing apart the misleading noise (deceptive characters). Whereas your system, Carl, permitted us to organize and communicate our classifications, it was Hennig's philosophy that finally provided the theoretical fabric that wove the 'entities' of Nature together, what students later came to understand as *taxa*

Centropristis philadelphica (L.) (Serranidae), the Rock Sea Bass or Cherne, was first placed by Linnaeus in the genus *Perca*

(hypotheses of kinship, natural affinity). With a quiver of paternal pride, I am pleased to report to you, dear Carl, that ichthyologists were in many ways at the forefront of this 'methodological insurgency', even though, lamentably, the early fires of revolution seem to have prematurely abated.

In summary, Carl, I can relate to you that modern systematics still finds itself suffering from growing pains, in which predominant trends are for combining disparate sources of data (from morphological to molecular to ecological), even though in many instances this seems only to cloud the true evolutionary signals we seek, not clarify them; statistical, model-based phylogenetic inference has even managed to gain the upper hand and is routinely implemented alongside, or even in lieu of, parsimony analyses. There are even those who would, foolhardily, discard your classification system as logically inconsistent with the patterns of evolution (the phylocoders). These are but signs of modern times, wherein even alleged quick-fixes such as barcoding and automation have, perhaps unsurprisingly, found a niche, leaving us to question: how is it possible, in science, for superficiality to be so in vogue? For propaganda to outfox common sense?

Somehow, it seems we have come full circle, as what is presently needed in taxonomy and systematics is a 'return to basics'. In other words, professional taxonomists and comparative biologists must venture to *add robust hypotheses* to the general reference system initiated by 18th-century naturalists, among which you were a leader. But we must do this by getting back to a serious concern over the *quality* of data and its detailed description, thus augmenting comparative biology and its long-earned conceptual independence. There is still so much to do, my dear Carl; describing millions of new species is but one part of the task at hand—and really just an initial step—as naturalists must unravel their relationships and shed light on the historical processes that have led to their current distributions. (You may not wish to hear this, but good old Georges-Louis Leclerc, Comte de Buffon, your Nemesis, who disparaged your system of classification as artificial, "imposing" and presumptuous, was also very much on the right path in his own earnest endeavor, even though you regarded him as a "God-abandoning transformationist".)

In this perspective, Carl, your contribution—and Buffon's, too—are as relevant today as ever, as we continue to search for the natural kinds of fishes through the study of their character systems, with many lingering problems to solve, and all the while still marveling at their superlative diversity. So let us rejoice in the splendid longevity of your composition!

With much yearning, I remain devoted,

Petrus

Marcelo de Carvalho (for Petrus Artedi [1705–1735], a founding father of modern ichthyology)

Departamento de Zoologia
Instituto de Biociências
Universidade de São Paulo
São Paulo

Välborne Herr Archiater
    Nådige Herre!

För någon tid sedan hade jag äran gifva Herr Archiater wid handen, att jag war ganska nära att blifwa
engagerad för Africa och nu för tiden i lika ju
mistkhet namna att affairen är afgjord, och jag
så wida resferdig att jag är nögd gå tilsjös
hwilken dag dess för hälst Capitainen behagar.
  En af Mina wänner, som jag jedst hade omtala
wid namn Smeathman reste för mer an ett år
sedan til Custen af Guinea i Africa för att sam-
la Naturalier, och har där warit så lycklig
att Hrr Subscribenterne ej kunnat annat an wil-
fara hans begäran, och med andra behofwer
äfwen låta honom få ett Systeme och en Syste-
maticus öfwerskickade, jyckan har och wärit
dem så benägen, att de fått det förra af bästa edi-
tionen men hwad den senare angår eller Mr. Sys-
tematicus så måste Hrne låta sig nöja med en
af förstra uplagan, jag har blifwit föreslagen att wara
asistant till Hr. Smeathman, men har stor
orsak att befara jag ej är wuxen en sådan
charge, serdeles som det wäntas, att jag skal be-
skrifva hans samlingar, utan att jag har
den insikt som dertil fordras, och utan att
Hrne weta om jag kan åtskilja fisk och fogl
emedlertid hoppas jag mig ej blifva aldeles o-
nyttig, ty jag kan ofta wara ganska idog,
men det hoppas jag wenner ursäkt att jag
måste utellemna och förbiga sådane observatio-
ner som erfordra större hastigt och försåkenhet
än min hvilken ej wäger mycket.

Den Öen wi komma att bebo är kallad Bananöes och ligger wid utloppet af Sierra de Sion, en flod ofwa, så angenäm som landet hwilket den genomskiär är löftigrande, marken är derfruktbar, luften sund, nödwändiga behöfver til öfverflöd; inwånarne Angelymar ifrån hwilka jag nödigt skiljes om de äro sådane som de ras anhörige i London. Ödemarker, wildjur och oförmodade siukdomar äro obehageliga föremål för min tankegåfva. Men Herr Archiatern må wara försäkrad att jag ej gör mig verre tanka om Africa än Halla Lappmark, jag wet at jag är mistagen, men jag hoppas ändå med Guds nåd att åter få se Fädernes bygden och först Musernas boning så jag med förnöjelse för lemna Herr Archiatern sål så wäl, af vidrige som lyckligare Öden men först ett måste jag först anhålla om: har Herr Archiatern ännu någon ömhet för den minsta af Propheterna så låt hans (kanske sidsta) begäran winna bifall, jes är att Mr Smeathman måtte blifva hedrad med Herr Archiaterns Correspondence, hvilket jag wet han på det högsta skulle til önska sig, och om jag ej är bedragen säkert igen känner med bättre kraft än löften, hvilka andra wäl gifva men aldrig tanka att upfylla. Hvad mig angår, så skulle

Jag väl önska att få en plan eller instruction att rätta mig efter; men som Jag wet att Herr archiaterns tid är för myc=ket uptagen med wicktigare göromål, wil Jag ej derow entrådet avhålla, varandes fuller kommen nogd om min begäran, för Mr Smeathman winner lika benäget bifall som den för Mr Miller fordom.

Sedan Hr Banks och Dr Solander kom= tilbaka ifrån Island har den sednare blif= vit eftertradare efter Dr Mathew vid Mu= seum såsom 2dra Bibliothecarius, och då Dr Matthew fått Salig Dr Knights Tjenste i händelse att Hnk rest til foversion tror ej att Dr Solander welat emottaga slegt tellbs

Skulle Jag ej mer få folka mina tankar för Herr Archiatern så önskar Jag nåd och frid af Högden och rik wälsignelse öfver des Familia til sednaste led, ack! måtte den sent sakna den bästa af fäder, Wettenskapen des stöd och jt den säkraste och stötte befordrare, är des önskan hvarmed Jag förblifver

Välborne Herr Archiaterns
min nådige Herres

London d 12 Jan.
1773.

ödmjukaste
tienare
And Berlin

Anders Berlin (1746–1773) was one of Linnaeus' students who travelled to West Africa to collect plants with the Englishman Henry Smeathman. Berlin was one of the unfortunate "apostles" who never returned, he died of a stomach illness on the Iles de Los off the coast of Equatorial Guinea only a few months after he arrived in Africa. He humbly writes to Linnaeus from London on 12 January 1773.

"Some time ago I had the honour of informing you Doctor that I was pretty close to be engaged in going to Africa and now I can humbly say that the business is done and I am ready to leave any day the captain cares to name. One of my friends, whom I have already had the honour of mentioning, by the name of Smeathman, travelled more than a year ago to the coast of Guinea to collect natural history specimens and has been so successful there that the gentleman subscribers have not been able to do other than to grant his request and together with other requirements also let him have an assistant sent out . . . I have been proposed as an assistant to Mr Smeathman but I have great cause to fear that I am not equal to such a charge, as it is expected that I should describe his collections without having the insight necessary and without the gentlemen knowing if I can tell the difference between a fish and a bird, however I do not think that I would be totally useless, as I can often be rather industrious and with that I hope I will win an excuse that I will have to leave out and pass over such observations which require more insight and experience than mine which does not carry much weight. . .

The island we will live on is called Bananas and is situated in the estuary of Sierra de Lion, a river which is as pleasant as the country it cuts through. The soil is fertile, the air healthy, everything is in abundance and inhabitants Englishmen, from whom I will part if necessary if they are like their relations in London. Deserted areas, wild animals and unexpected illnesses are disagreeable thoughts, but you Doctor must rest assured that I am not making myself think any worse of Africa than of Luleå Lappmark. I know that I am mistaken but I hope yet by the grace of God to see the home of my forefathers and first the home of the Muses when I can, with pleasure, tell you Doctor about horrible as well as happy events, but first there is one thing I have to ask you, if you have any affection for the smallest of the Prophets, then let his (maybe last) request be accepted, which is that Mr Smeathman may be honoured with your correspondence which I know he would sincerely wish . . . As far as I am concerned I would like to receive a plan or instructions to follow, but as I know that you Doctor are very busy with more important matters I do not want to stress this too much.

Should I not anymore be able to tell my thoughts to you Doctor then I wish you the highest peace and mercy and blessing for all the family and may it be a long time before they have to miss the best of fathers, science its support and I the safest and best supporter, this is the wish with which I remain,

My Wellborn Doctor's, My Gracious Lord, Your Most Humble Servant,

Anders Berlin"

ROYAL BOTANIC GARDEN EDINBURGH

25 August 2008

Our Ref: SB/CL0001

Carl Linnaeus
Uppsala
Sweden

Dear Carl

I am sorry that you have not heard from the Royal Botanic Garden Edinburgh for some time. Your last letter to us was to Professor John Hope and I am replying in his stead. This response is long overdue, but we have all been rather busy trying to document plant species and there just never seems to be a spare moment for simply keeping up with old friends. I am sure you know how it is. Frankly, this global warming business is not helping either. As if nature was not already on the retreat, what with more than six billion of us swarming all over the planet. Of course, there are nine times as many of us humans today as there were when you were a young man and that puts a lot of pressure on the poor old planet. You would think that with all those billions of people we would be able to recruit enough taxonomists to press on with the task you started: cataloguing life on Earth. But as you know, that is just not seen as a priority these days when apparently there are more important matters to attend to and so many new ways of keeping scientists busy. Of course, we botanists know that it is the most important matter we need to attend to but nobody seems to be listening. So now that all of this environmental change is going on it seems that those few of us there are will just have to work harder than ever. But if nature is on the move and will not even stand still while we document it, what are we to do? If anyone has the answer, I am sure it will be you.

I should really introduce myself since we have not had the good fortune of meeting. You knew some of my predecessors at the Garden and, of course, you knew John Hope rather well. He was a very enthusiastic teacher of your Sexual System and encouraged many others to adopt it here in the British Isles. Hope was such an admirer of yours that he commissioned one of the finest architects in Scotland, Robert Adam, to build a fine memorial to you. We brought it with us when we moved to Inverleith from Leith Walk and it is still much admired today.

I ought also to bring you up to date with what has been going on in Edinburgh since your last correspondence. There have been quite a few

Regius Keepers since John Hope, 11 of them, in fact. But we certainly have not forgotten either his correspondence and exchange of seeds with you or his important contribution to the Scottish Enlightenment. In fact a new building named in his honour will be opening in Edinburgh next year. The John Hope Gateway, as well as providing a warm and informative welcome for visitors to the Botanic Garden, will be a place for people to find out more about the natural world and how it is changing. There is so much confusion about what, if anything, people and their families can do to stop the warming of the planet and the destruction of nature. This new building will be a place where they can come to learn about and to discuss these difficult challenges. Visitors will then be able to go out into the Garden and enjoy everything it has to offer, from the tens of thousands of plant species we are growing to Hope's monument to you.

I wanted to let you know that we still use as our institutional emblem, or logo as such things are known these days, that delightful little plant, *Sibbaldia procumbens* L., that you so kindly named in honour of Sir Robert Sibbald, one of the founders of the Royal Botanic Garden Edinburgh. It is illustrated in a slightly stylised way but hopefully still quite recognisable. I wonder if you would mind looking at it for accuracy? We will be happy to modify it in any detail, if you think that would be an improvement. *Sibbaldia* is a widespread snowbed plant and already this year I have seen it flowering well on several Scottish hills including Stob Ban, Beinn a' Chreachain, Braeriach and Stob Coir' an Albannaich. However, I have noticed that it is only found in high places where the snow lies for several months in winter. I also wanted to ask your advice on the likely future of this particular wildflower. It concerns me that it may be lost to Scotland if the climate changes as we understand it might. I know that it is a widespread species and will surely find high enough mountains in other lands but here we are proud of our natural heritage and do not want to see it slipping away. What are we to do?

Your advice, as always, will guide our best endeavours. Hope you are keeping well.

Yours sincerely

Professor Stephen Blackmore FLS, FRSE
Regius Keeper

A specimen of *Sibbaldia procumbens* L. (Rosaceae) from Linnaeus' herbarium
(LINN 401.1)

Carl Linnaeus

Sir London 26 March 1745
Printed in Linn. Corresp.
v. 2. 440.

On board the *assurance* Capt Fisher is a Case of American plants in Earth; They are a present to you from my good friend Dr Lawson. knowing this his intention by his consulting me to know what plants I thought would be acceptable, I selected these as being hardy and Naturallised to our Climate & consequently somewhat better adapted to endure your colder Air, yet I wish they do not require as much protection from the severity of your winters, as plants from between the Tropicks do with us / possibly you have already some of them, yet if but a few of them be acceptable, I shall be much pleased, And whatever other American plants in this inclosed Catalogue will be acceptable, you may freely command any that I am possessed of In regard of that Esteem your merit claims

I am Sir

This Case of plants were intended to be sent last May and they were sent to the ships with the consent of the Shipper yet they were refused to be taken on board

Your most Obedient Humble Servant
M. Catesby

Cypressus Americanus
Arbor Tulipifera
Cornus Americ:
Populus Nigra Carol:
Periclymenum
Barba jovis Arborescens
Phaseoloides
Arbor Virg: Citric: folio
Euonimus Americ:
Aster Americ: frutescens.

Lychnidea flore purpureo
Itrex Aceris folio
Bignonia —— Catalpa.
Angelica Spinosa
Pseudo Acacia
Anapodophyllon Canadense
Phaseoloides frutescens
Rubus Americanus

Mark Catesby (1683–1749) collected in what are now the southeastern states of the United States and in the Bahamas. He sent specimens to prominent English botanists, but not to Linnaeus himself. His book *The Natural History of Carolina, Florida and the Bahama Islands* (1730–1747) was one of the first to catalogue the diversity of this region. He writes to Linnaeus sending a shipment of live plants from the Americas he has selected as a gift from Dr Lawson; Catesby chose plants acclimatised to England, thus better able to withstand the colder climate of Sweden. He adds a list and offers more, saying "Whatever other American plants in this enclosed Catalogue will be acceptable you may freely command any that I am possessed of—In regard of that Esteem your merit claims".

30　*Letters* to *Linnaeus*

Illustration of the Eastern Chipmunk of North America—*Tamias striatus* (L.)—and the "Mastic Tree" of the Bahamas—*Sideroxylon foetidissimum* Jacq. (Sapotaceae) from Catesby's *The Natural History of Carolina, Florida and the Bahama Islands* (1730–1747). In his illustrations, Catesby often juxtaposed plants and animals that did not occur in the same region.

Dear Linnaeus

You are in Heaven. I am in Hell. I should consider myself fortunate that the authorities here are allowing me this one privilege to write to you. No doubt it is because of your esteemed standing and the 250th anniversary of the 10th edition of your *Systema Naturae*, which this letter honors.

How did your friend, Anders Celsius, end up in Hell? Rank insubordination. As you know, I died in the spring of 1744. When I came before Saint Peter, I fully expected to be ushered into Heaven. After all, two years earlier, in the service of the Almighty, I had described the Centigrade thermometer at the gathering of the Swedish Academy of Sciences.

Saint Peter went through his spiel—the bounties of Heaven, the brutality of Hell. He made a special point of telling me that the temperature in Hell was 800 degrees Fahrenheit.

"FAHRENHEIT!" I exploded at him. *"You are measuring the temperature of hell in Fahrenheit! Not Centigrade? Imbeciles! Nincompoops! The holy scriptures advertise you as all knowing. Hah! The Fahrenheit scale is vulgar, an insult to numerical grace—imagine water boiling at 212 degrees and freezing at 32. My thermometer scale has mathematical elegance: water boils at 0 degrees and freezes at 100 degrees."*

What can I say, Carolus. Saint Peter doesn't suffer insubordination gladly. He tut-tutted and tossed me into the underworld with the rest of the rogues. But not before he told me—with much glee, I might add—that none other than *you*, my dear Linnaeus, betrayed me after my death by reversing my Centigrade scale. Now water boils at 100 degrees and freezes at 0 degrees. Saint Peter even gave me the citation: *Hortus Upsaliensis*, 16 December, 1745, in which you use the "forward centigrade scale" to describe the temperature inside the orangery at the Botanical Garden of Uppsala University. Bah! How would you have liked it if I had summarily reversed your precious taxonomy, say, moving *Homo sapiens* from your Order Primates to your Order Bruta?

I imagine I should really be grateful. After all, you were not the only one to reverse my temperature scale. Quite a few of our colleagues independently did so as well: Pehr Elvius, the secretary of the Royal Swedish Academy of Sciences; Christian of Lyons; Daniel Ekström; and my own student, Mårten Strömer. But you, Carolus, were the first.

So, I must acknowledge, in hindsight and in humility, what few people realize. It was you who led the way in making my temperature scale a worldwide standard. Celsius is ubiquitous—in weather broadcasts, in scholarly journals, in thermometers. You made Celsius a household name, surpassing your own. References to "Linnaeus" and "L." pervade taxonomic treatments, but I dare say that "Celsius" and "degrees C" are omnipresent. My body might be in Hell, Carolus, but you promoted my name to Heaven.

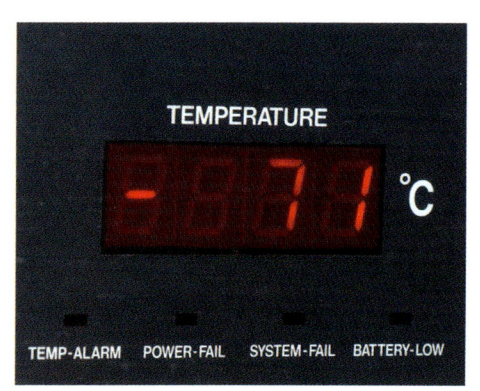

The Celsius thermometer on a −70 degree C freezer for storing DNA samples

Which brings me to this anniversary of your *Systema Naturae*. I do not take pleasure in imagining that you must be suffering a personal hell. You trusted that after 250 years the census of life on Earth would be complete. Yet fewer than two million of the estimated 20 million or more species of plants, animals and microbes have been discovered, documented and described. This record is shameful. Worse, it is irresponsible, ethically and scientifically. Extinction is outpacing taxonomy. Perhaps 50 years remain to discover the life of the planet before it is depauperate enough to cede to paleontology. Unfortunately, as in 1758, taxonomy suffers from too few experts and too little money. But a third problem is that 250 years after you systematized taxonomy, it is still a cottage industry, still being done the old fashioned way.

Were you alive, I imagine you would lead the transformation of taxonomy to industrial strength. You would automate major components of species discovery, documentation and description, warping the pace of taxonomy from years to hours. The genomics and neuroscience enterprises are well along this road for genes and brains. Industrial strength machines elicit, capture and warehouse reams of raw data—incomprehensible petabytes for a single person; then, massively paralleled computer systems make sense of the data, detecting patterns, turning information into knowledge.

As you know, Carolus, all the tools—analytical, artificial intelligence, informatics—are available to taxonomy for rapid, assembly-line species identification. They can automate the capture, coding, identification and visualization of gene sequences and 3-D morphological information from every plant and animal specimen collected by inventory expeditions. They can automate genotype-phenotype association and species recognition, allowing the taxonomists to apply their enormous expertise to the truly novel.

And Google's MapReduce—I'm sure Heaven has Google—is perfect for squeezing knowledge from colossal amounts of text data across a server farm of computers. In the past 250 years, your followers have created a Babel of individual species names for millions of specimens in museums and herbaria. Perhaps some taxonomic Google can finally wrestle order from the names and the plethora of concepts taxonomists have for each particular species.

I cannot imagine how you will lead this revolution from Heaven, but knowing you, Carolus, you will succeed.

With admiration and hope, I remain your friend.

Anders Celsius.

Olof Arenius' (1701–66) painting of Anders Celsius, from the Uppsala Astronomical Observatory

Postscript. You need not take the time to answer this missive—we are not permitted to receive letters in Hell. Indeed, Hell has no paper—save for these three sheets of letterhead from a Biodiversity Institute at the University of Kansas. Its former director, a paleontologist named Krishtalka, somehow managed to smuggle them in here. He will not tell me why he was sentenced to Hell. But, of course he knows of you, and he shared with me his outrage at your classification of Canadians in *Systema Naturae* as "*Homo monstrosus*, head flattened". As you might imagine, he is a Canadian.

L Krishtalka

Linné, kom hem!

X-Sieve: CMU Sieve 2.3
Date: Fri, 2 Sept 2008 10:47:17 -0400 (EDT)
Subject: Linnaean classification under threat: urgent action needed!
From: "Joel Cracraft" <jlc@amnh.org>
To: "Carl Linnaeus" <c.linnaeus/Fellows.18thcent@taxonomisthvn.org>

Professor Carl Linnaeus
President, Taxonomists/Biology/Heaven

Dear Professor Linnaeus:

Oh miracle! I have finally located you, as the message did not bounce back. Please forgive this intrusion on your time, my dear Professor, as it is well known you are a very busy person. It has been extremely difficult to find your address after all these years, and I tried many of your closest followers down here to no avail. In the end, it was your colleagues in the 20th century subsection, Willi [Hennig] and Ernst [Mayr], who kindly forwarded your contact information.

I am writing only because there is a scientific emergency here that requires your immediate intervention. It will no doubt be frowned upon by your superiors to become involved in such matters, but once this crisis is explained, I am sure they will endorse action to solve this problem. After all, it involves an impending disruption to the natural order of things here on Earth; and frankly, it truly requires a response beyond anything I can imagine happening here.

For the last 60 years or so, your taxonomic system—and potentially your legacy—has been under siege, and the situation can only be said to be getting worse. I know 60 years is a trifle in the Grand Scheme of things, but for those of us still packing our bags, recent events have become more and more distressing. You obviously know that bandwagons in taxonomy have come and gone. (Some like the Quinarians were discarded quite quickly—speaking of those pentamerous fellows, would you please pass on my best wishes to the ornithologist Bill Swainson? I trust he has kept up with the literature and will know me.) But I digress and beg your indulgence. As I was saying, band-wagons have come and gone, but for some of us who revere your legacy, recent ones are truly disturbing. Charles Darwin, of course, was not entirely helpful to you, with his blatant attack on the fixity of species, but even he understood that your system was the only way to go and that it was easily amenable to his ideas. And I must say you have not been treated fairly on this evolution thing, as you clearly recognized that

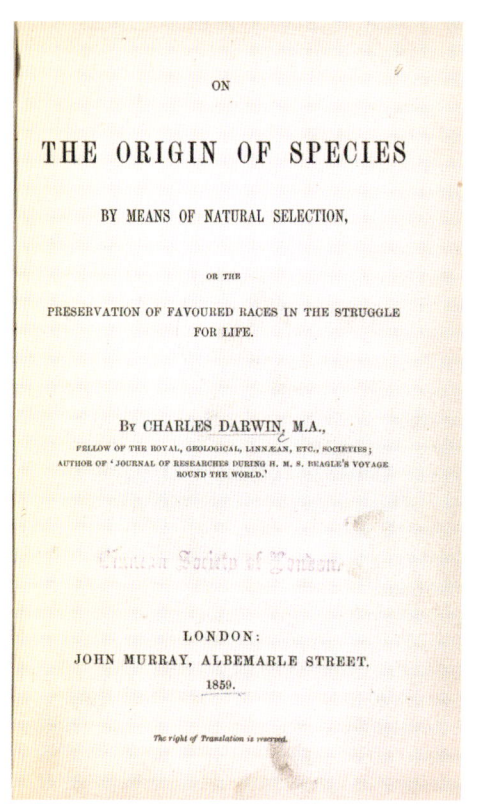

Title page of Charles Darwin's *On the Origin of Species by Means of Natural Selection* (1859)

new forms did arise over time. So no harm done. Such was the great power of your innovative thinking, dear Professor. Anyway, the first band-wagon to present a challenge emerged from Willi Hennig's monumental contribution in which even he, and some of his acolytes, had doubts that your system would work as more and more of life's hierarchy was revealed. Some of his followers, however, argued that such pessimism was unfounded and proffered there were many clever ways to accommodate increasing knowledge about life's natural hierarchy and its expression within the system that you created.

But now your legacy faces its greatest challenge, and much of it is coming from botanists who, of all people, should know better than to disrespect the clarity of your vision. Do they not remember that everything they learned about life's diversity started from their first exposure to your genius? Names! Characters! Hierarchy! System and communication! And species that mean something! How the hell (oops, sorry) could they be so blind to the fountain of all their knowledge? Do they actually believe their 'brilliant' ideas will resonate with young students, as indeed yours have all through these years? Incomprehensible arrogance, I say! You probably won't believe this, but even as they celebrate your life, some taxonomists—for Linnaeus' sake!—think we should dispense with binomials! That we should reserve naming for only those new species deemed to have some special purpose, and should IGNORE naming the others! Taxa with no names?

Some of these poor souls make the point that ranks of taxa are themselves a problem and give one a false sense of security, that taxa of the same rank are not equal, that they cannot be counted, and so forth. Well, duh! Such ideas are long-gone except perhaps for a few paleobiologists, ecologists, and misguided systematists who still have not been innoculated against evolutionary taxonomy and just love to count things. So most of us taxonomists are beyond arguing this point, and see names of higher taxa as being 'rank-free', or perhaps better, as 'rank-independent', but is this agreement a reason, a justification, to throw all hierarchy out the taxonomic window, that subordinate ranks themselves carry no information? I ask you, what are these %^&#$ thinking? Sorry for my intemperance, but this is getting too much to take.

Professor, all of this is just the beginning. I am just getting started! How about family names within other family names, within others? I swear, they are doing this! How about dumping taxonomic hierarchy? System? These misguided souls are so fixated with doing away with ranks, and replacing established names with their own, 'rank-free' names. I'll tell you, it is a feeding frenzy down here to get one's moniker after a spankin' new rank-free name (ironic isn't it). But then, once named, they stop. Yup, just like that. I know, you are asking 'where's the beef'? Where is the system? The hierarchy? Well Professor, if you listen to them, then sometime in the future a computer will take care of everything, solve all the problems, just like that, with a few lines of code. That old box of wires will take all the new names for the nodes on their very 'special' tree, and sort those names into a list—Professor, please keep calm, I'm just the messenger here—in order that one can 'see' how one group name fits relative to all the others. I know, I know, the false sense of hope that a computer can take care of everything, but, Professor, I have no doubt it can be done. One can easily imagine it, and I can give you an example—unimaginably, dear Professor, this is REALLY TRUE: "human, the Apiidae are in the Campanulidae, which are in the Gentianidae, which are in the Asteridae, which are in the Gunneridae." JAG svära till GUD den här er Sann! Computern über alles!

Names! Names! Names! Only names as far as the eye can see? Oh, where is the beauty of system, of hierarchy? We so dearly miss the poetry in your words! And, I might add, the ability to communicate, which you perceptively saw as being the *raison d'être* of all this.

Professor, this heretical thinking is gathering steam because a small number of taxonomists are beating their students into submission and drafting legislation to make their ideas the law of taxonomy. The old

Gentiana sino-ornata Balf.f. (Gentianaceae) from meadows in southwestern China

saying that students debated all viewpoints and then adopted those of their major professor has never been truer. Now I know you were not into many debates with your own students—and that certainly worked out for you!—but as we are reminded daily, things must change. Thus, if an intervention by you could encourage some of these students to push back with some common sense, maybe we can make some headway.

Professor, once again we have need for leadership, and there is no one more brilliant, more qualified than you to take this on. Certainly given your current position and powers, any action would have to be done by stealth, lest the whole scientific establishment think we taxonomists are more sanctimonious than we really are. What would be the best plan? You know, I bet you have already constructed a new comprehensive system to accommodate all those species in your garden, and that it is bigger and better than ever. What if we sprung that on the heretics? They would be humbled beyond belief and would have to back off!! Victory for your legacy would be manifest!! Do you think it would be possible to share your new system with the rest of us? If you are worried about security over these lines of communication, we could meet discretely, perhaps on a mountain top where you could pass on a disk? That's the conventional way you do these things, isn't it? In my case, at least, I promise to see that this new system will be slipped into the literature anonymously. But whatever you decide is fine by me. By the way, is the rumor true that you interjected phenetics here on Earth as a way to punish followers of the 'new' systematics and set the stage for cladistics? If so, we need another surreptitious intervention just like that, at this very moment.

Honorable Sir, please think about assisting your brothers in arms in this time of need. The simplicity, elegance, and innovation of your work are timeless, but time is growing short. Please give my regards to Willi and Ernst.

Hälsa Linnaeus, mästare av system och natur!

Your faithful servant,

Joel Cracraft

Joel Cracraft

X-Sieve: CMU Sieve 2.3
Stardate: -314324.0199036127 (STD)
Subject: RE: Classification under threat: urgent action needed!
From: "Carl Linnaeus"
<c.linnaeus/Fellows.18thcent@taxonomisthvn.org>
To: "Joel Cracraft" <jlc@amnh.org>

My dear Professor Cracraft:

Tell me something I don't already know! Please, not to worry. The 'heretics', as you call them, eventually will pay a price (the Fellows list here is quite short, and competitive, and there is an extremely lengthy process of vetting). Keep up the good work. But under no circumstances are you to attempt to contact me again. I have instructed Willi and Ernst in no uncertain terms that they are not to give out my contact information ever again to anyone. With respect to the misguided souls you describe as "still packing their bags", there are alternative places where they might be housed. (They could, for example, spend eternity with ecologists trying to do taxonomy. That would drive them nuts! Just kidding, but I might have my staff work on this.) Willi and Ernst are not yet Full Fellows, which will take at least seven or eight more decades of vetting, therefore they know the consequences of indiscretions. By the way, only one of them sends his regards.

Very truly yours,

Carl Linnaeus

Carl Linnaeus
President
Taxonomists/Biology/Heaven

Carl Linnæus

Dear Carl;

I have been meaning to write for some time, but I wanted to be in touch now, to congratulate you on your 300th birthday in 2007 and to give you an update on the project you started over 250 years ago to document all the species of the world, beginning with plants. Your *Species Plantarum*, which is still a standard reference, was a really monumental effort. But over the years, the need for some kind of new and updated equivalent has grown steadily more obvious. It is amazing that after all this time, and after all the work of many thousands of intrepid plant collectors who have followed in the footsteps of your students, many new species of plant continue to be discovered and described. I remember you once commenting that you thought there might be as many as 10,000 species of plants in the world. We now know that the actual number is probably closer to 400,000, with perhaps about 350,000 having already been recognized. A few years ago we set ourselves the goal of developing a global checklist of plant species, and we are making progress, but pulling together a new *Species Plantarum* is a really big job.

The western regions of the United States were not explored until long after Linnaeus' time (Sangre de Cristo Mountains, New Mexico)

On the other hand, in many ways, compiling a more or less complete list of the world's plant species, as well as basic information on their distribution and main characteristics, should not be quite as daunting as it seems. In some respects, it should be easier for us than it was for you. For one thing, the science of plant diversity that you started, now has many more active practitioners than in your day, and they are spread all over the world, rather than being concentrated in northwest Europe. We also have automated means of assembling, compiling and tracking published information on plant species. This technology should make the task very much easier. But on the other hand, with so much published research having accumulated, and in a rather haphazard way over the years, we risk spending more time reading what others have written rather than looking at the plants themselves. And with so many people now involved, all working in different circumstances with different priorities, it is also hard to develop the kind of single-minded focus needed to deliver a new *Species Plantarum* in a reasonable period of time. Often, it seems difficult to muster the necessary shared sense of purpose. That's a real impediment, because these days it really isn't possible for one person, or even one organization, to do the job almost alone, as you once did.

But on top of these difficulties there are a couple of structural problems built into the fabric of modern work on plant diversity that are not so easy to overcome. Most fundamentally, there is a real tension between the further development of plant diversity science as a modern scientific discipline and a more utilitarian approach that focuses on delivering the kind of useful synthesis that non-specialists and other scientists really need. I know that both these objectives were dear to your heart. Somehow we need to figure out how to do both, just as you did. On the one hand understanding the diversity of plant life is a legitimate and important science in its own right. But on the other hand we also need to deliver the kind of comprehensive syntheses that will be really useful to the many people who are not specialists, but who for one reason or another deal with plants in their work and everyday lives.

And while I'm mentioning problems, there is another built-in difficulty with modern work on plant diversity. I'm sorry to bring this up. It is a slightly uncomfortable issue. But I have to say that in some respects it was not a great idea to make the name of every species a binomial. It started out as useful shorthand for you but the idea has caught on in a really big way. It is now the standard approach across the whole of biology. We are all so used to it that we take it for granted. Over the years we've come up with some wonderfully memorable names. But while in some ways the binomial system has served us well, it has also turned out to be a bit of a burden.

We have always had to make judgments about what should be included within a particular species. But the use of binomials, which we are now locked into, means that to decide on the most appropriate name we have to judge not just the limits of species, but also the limits of genera. Using binomials means that concepts of both species and genus identity are built into the name. This is really not helpful and has contributed to some appalling and very time consuming problems with synonymy. I know that you faced similar problems in your time too, with multiple names for what is obviously the same species. That will always happen. But having to argue about the limits of genera as well as about the limits of species introduces another layer of complexity. Most of the species level entities that you named in *Species Plantarum* are still recognized exactly as you had them, but many of their names have been changed (sometimes several times) solely because subsequent specialists have concluded that the species belongs in a different genus.

Instability in the names of plants is bad for us and for our users. The built-in problem with binomials is not the only cause of this instability but it is a big contributor to a problem that has really got out of hand. More than a million names have now been published: probably about three times more than the actual number of species. The wonderful tidying up of this kind of mess that you managed to achieve in *Species Plantarum* has really come unraveled over the past couple of hundred years.

I wouldn't feel so frustrated about all this if it wasn't for the fact that we specialists in plant diversity now have more important things to do than argue about the synonymy of names. It doesn't help our science and it doesn't improve the important services we need to provide. Increasingly we are in the front line of efforts to ensure that plant species in many parts of the world don't disappear completely. And if we don't work to ensure that plant diversity is maintained then who will? It is a privilege to able to devote our working lives to understanding the variety of plant life. But with that privilege also come responsibilities. Already there is clear evidence of plant species that have gone extinct, others survive only in cultivation, and many, many more are heading in the same direction. The world has changed, almost beyond recognition, since your day. Nowhere on the planet is free from human influence and you would not believe the scale of current human impact on the global environment.

Franklinia alatamaha Marshall (Theaceae) is known only from plants in cultivation

Well, I'm sorry to go on (and I'm especially sorry to have brought up the delicate issue of binomials) but it feels good to get all this off my chest. The main point though is to send you my best wishes, and to thank you for everything that you have done to stimulate the scientific study of the global diversity of plant life. Many have followed in your footsteps, but none have made a bigger or more lasting contribution. I hope that in the coming years we can muster the focus necessary to produce new and useful works of synthesis that measure up to the standards that you established.

Wishing you well in your continuing retirement.

With my very best wishes

Peter

Dear Dr Linné,

It seems strange to be writing to you at this distance in time, but knowing your exceptional interest in the natural world, I thought you might be curious to know about equipment and methods available to us in the 21st century that excite me (among many others) with regard to the study of plants. Of course, the methods available to botanists now are equally available to zoologists—but that is not my prime concern here.

At the outset, I am a little diffident in approaching you on the subject of plant anatomy, which relies heavily on the use of illustrations and images in its application. This is because I realise that you expressed an aversion to drawings of plants, regarding written descriptions to be of much more value; also, in order to understand what you are seeing as a plant anatomist, reasonable drawing skills are valuable. Even in this respect, we now have processes for obtaining excellent illustrations that do not require any manual artistic skills!

Although reasonable degrees of optical magnification were available to you, you seem to have preferred to use simple lenses to examine flowers, for example. Indeed, the study of micromorphology beyond floral parts and larger hairs does not seem to have been of great concern to you. At higher magnifications a new world opens up! I have wondered what might have happened to the course of taxonomy had you become enthralled by microscopy at an early stage in your career. Perhaps you would have still wanted to order nature, but you might easily have become diverted! Who knows? However, my experience of taxonomists today would suggest that a few would always be satisfied with such morphological information as can be obtained with a hand lens. Fortunately, over the 44 years I have worked at Kew, such people have become rare, if not extinct! Without doubt you would be in the forefront of those using the new armoury of methods if you were alive now.

By using optical microscopes we have been enabled to look at the surfaces of plants in much greater detail, and our ability to make high quality thin sections of leaves, stems, roots and flowers for microscopic examination has extended considerably the range of diagnostic and taxonomically useful

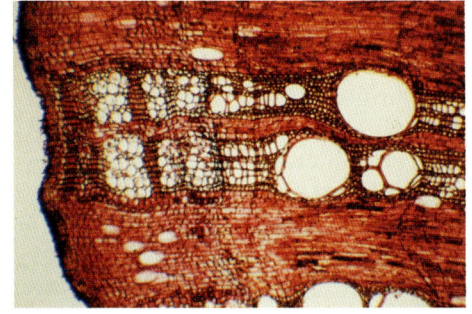

Cross section of part of a grape stem (*Vitis vinifera* L., Vitaceae)

characters. The use of selective dyes or stains on the sections or surfaces that are mounted on glass 'slides' has enabled us to distinguish between different tissues, and by using other techniques, find out their function. More complex modifications to the optics of microscopes beyond bright field microscopy to dark field, phase contrast and the like have extended their use. There are other instruments available to us that enable us to visualise the three dimensional structure of internal parts of plants without actually destroying them, such as confocal laser microscopes, which use light in a modified form.

Your system of convenience, using the numbers of floral parts to divide plants into groups undoubtedly made widely and readily accessible the pleasures of studying and identifying specimens. Many keen amateurs were led to become expert in such matters through this enlightened scheme. Nowadays there is a general interest in systematics that indicates natural relationships through phylogeny (as I am sure others who have written to you recently will have pointed out). Anatomical characters have been used successfully in this process. However, we still produce artificial keys that help people to identify items. For example, wood used in furniture, construction and other artefacts often can be identified using artificial keys with characters observed from the microscopic structure. We also rely heavily on the written text in combination with illustrations to help people identify fragments of plant material that may be recovered from ancient archaeological sites—or, perhaps of more immediate interest to you—to try and identify fragments of plants that have possibly poisoned people.

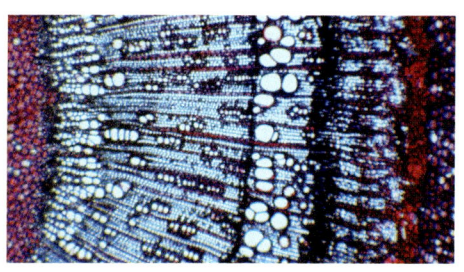

Cross section of part of an ash (*Fraxinus excelsior* L., Oleaceae) twig

The inventiveness of people has continued to drive technology beyond the use of light to examine specimens! You will have found that some lenses were better than others at resolving fine details. It will seem improbable to you that even the physical properties of light are limiting in what can be resolved optically. I won't bore you with the details, but it has been found that light itself can be dispensed with, and microscopes have been developed using what are called electrons that give very much greater resolution, allowing us to delve into the fine structure of the actual workings of the plant. The first of these 'transmission electron microscopes' work on similar principles to light microscopes, and a very thin section of suitably stained tissue is substituted for the microscope slide. Electrons, rather than light are made to pass through them using lenses not of glass, but with 'electromagnetic' properties, to focus the electrons and produce an image. The images unfortunately cannot be viewed directly, but have to be captured for examination on special material.

When a student myself, I was fascinated by this equipment, but a wise professor suggested that such studies could readily divert me from the essential basic ground work in the comprehensive study of systematic plant anatomy that still needed to be done, and I listened to him. Only later in my career did I use another form of electron microscopy, that provided excellent three-dimensional images of plant surfaces, called scanning electron microscopy, to supplement and extend my studies. I am glad in many ways that you concentrated on the bigger picture, and laid the foundations of the scientific study of natural history, and were not side-tracked by the developing interest in optical microscopes.

I write to you in the unusual capacity of President of a Society that perpetuates your name, and looks after many of your specimens and books—The Linnean Society of London. You will, I am sure, be pleased to know that even 300 years after your birth and 250 years after your 10th edition of *Systema Naturae* you are highly respected, and that the interests in the natural world you kindled in so many of your contemporaries and students continue to flourish in the biological community, both professional and amateur. Many of your collections can now be seen by people from round the world, without them even setting foot out of their studies—but that is another story!

I am, sir, with deepest respect, one of your continuing admirers,

Yours sincerely,

David Cutler

Petiole of *Liquidambar styraciflua* L. (Altingiaceae, Sweet Gum) in cross section

Carl Linnæus

Dr. Carolus Linnaeus
Somewhere in Time

Your Excellency:

Would it be permissible for me to use a less formal address for this message? Acknowledging that you fully deserve the noun in the above address, and being aware of the nobility you embody, I like to think that your reputation for friendliness would smile on a move to tone it down a little in a letter from a great admirer and, I hope, friend.

Dear Dr. von Linné:

Now that I think of it, if we are friends it's possible you could favor me, out of my deep admiration for your sense of order and the evident benign concern your life's work has shown in dealing with all living things (and even in the kingdom that is not living)—that I might, without terror, address you as one who has been a friend to me for most of my life. In the belief this would be all right with you, I make bold to restart:

Dear Carl,

As one of many who are beneficiaries of the order you have brought to the complexity of life, I know of no way to properly thank you for this service except to express in this letter what it has meant to my life. When I was 13 years old, I saw for the first time the inside of a big city library. Stacks arranged in narrow rows and extending for great distances and multi-storied heights, displayed such a number of books that I was overwhelmed with the suffocating feeling that if I lived to be 500 years old I could not read them all.

Shelves of Linnaeus' books in the Linnean Society of London collections

I came to realize that there would be no point in reading them all, since there would be redundancy of subject and areas not necessary to the quality of my life. And I stumbled onto the fact that if there was some order in sifting through this myriad of publications, sorting them into categories and subjects and fleshing out a plan for reading, then I might be able to make the best use of my reading time and maximize the positive impact of such a project.

Among the scientific material I read were treatises on taxonomy, and this does for the world what my plan was doing for my reading, and I have you to thank for accomplishing this, and bringing order to the complex web of life on our planet.

How did you ever have time to become a medical doctor? And according to all accounts a very good one. Well, it is said "If you want something done, ask a busy man—idle men don't have the time." I don't need to ask you to do something for me, because you are doing this, all the time. My joy in viewing the tree of Life and knowing how and where various forms on it are placed, is ongoing, as new things are discovered, and as more old things come to light and take their proper places.

Again thanks. You need not answer this. But if you did, please dispense with "Mr. Downs" and call me "Hugh."

Sincerely,

Hugh Downs

Hugh Downs
Paradise Valley,
State of Arizona
United States of America
Northern hemisphere,
Planet Earth
Solar System
Milky Way Galaxy
Messier Local Galactic Cluster,
Coma Berenices Super Cluster,
Visible Universe

PS. Please forgive my detailing my location, but your work has made me mildly obsessed with taxonomic order. -H

Dear Carl,

I am taking this opportunity to congratulate you on the tremendous effect you have had on humanity. Your inspiration and hard work laid the groundwork for our modern understanding of the living world. Perhaps even greater a gift is that esteemed academics still debate the classification of a species. Furthermore with the advent of genome mapping we have opened a new door full of reasons for one classification or another. The fact remains that we have a taxonomy that we use as our bulletin board of understanding.

I mentioned to my wife Allison DuBois, who is a woman of advanced education in History and Political Science but is best known as an author and for her work with the police and as a Medium, about your truly revolutionary and inspired development of a system for naming plants and animals. In general she works with people who are no longer alive and specifically helps determine how and by whom this came to be.

She looked up from her book and asked, "What did he do?"

So I replied that you invented the modern method for classifying flowers, trees, vegetables, mammals, reptiles…in essence everything we consider to be alive.

She questioned, "He named everything?"

I said, "You know how there is a scientific name for a green bean?"

She needled me, "He claims to have named the green bean?"

At this point I knew that she was joking, but I replied, "No, he invented a taxonomy."

That's when the bell went off for her: "Now I remember, I do know a few big words. I did go to a university you know."

This is when I realized how influential your work truly is; it has given form to the formless. It is a framework for the rest of us to use in our chosen professions and is more complete that many realize. The inclusion of all that is known, whether observed life or, in a few cases, mythical beings means that your framework is truly scientific. It does not count out a theory merely because it has never been witnessed. It leaves the framework open for extraordinary evolutionary advances while allowing the taxonomy to be ever changing as a result of scientific debate.

With gratitude,

Joseph Dubois
Joseph DuBois

The Green Bean (*Phaseolus vulgaris* L., Fabaceae) from Linnaeus' herbarium (LINN 899.1)

Encyclopedia of Life

Most Noble Sir,

I write to you on the 250th anniversary of the publication of your definitive 10th edition of *Systema Naturae* (1758), a work that will shine for as long as men continue to value finding order in the natural world. Your life's work produced the means to record biodiversity, and to organize it into a hierarchy that is continually remodeled when new discoveries are made.

Much has changed since you studied the structure of life. You named 12,000 species of plants and animals, and when we refer to these, they are followed by the single letter 'L.' to denote the special attention they received from you. Using your binomial system, we have now named some 1.8 million species, and some taxonomists estimate that there could even be more than 100 million more species waiting to be discovered.

Today, the plants and animals are joined by many other taxa, including fungi, protists, archaea, eubacteria, and viruses. You suspected that bacteria, the "minute worms" as you called them, caused diseases, but they turn out to cause much else. During your century, the great discoveries in the field of chemistry showed us that nitrogen, oxygen, and carbon dioxide are the major atmospheric gases. But we now know that they are present in our air because of the constant activities of archaea and eubacteria, and furthermore, descendants of some of these cells are thought to have taken up residence inside the cells of animals, plants, fungi, and many other beings, where they provide both the fuel and the furnaces in which those organisms consume that fuel.

In your day, you were able to comprehend and organize the known diversity of the natural world. But today, not one of the major groups of organisms can be known by a single person. In fact, no book or collection of information on life forms currently compares with *Systema Naturae*, although we are in the process of creating such a compilation, one that will eventually represent every species in the entire world.

In your natural system, you identified genera as being the fixed forms in which nature is organized, and you believed that species were not fixed but could under some conditions blend through hybrids.

However, naturalists today appreciate that species are in most cases the fundamental units of evolution, the leaves of the tree of life.

'Tree of Life' might well sound to you like a poetic metaphor, and indeed, many naturalists have used it in such a way. But one naturalist, an Englishman named Charles Darwin, who lived for much of the 19th century, showed that the tree is not a metaphor. Rather, it is the process of evolution, of descent with modification, that leads to the tree of life and is the underlying reason your system of classification succeeded so brilliantly.

In addition to naming new species, many taxonomists today are interested in determining the order in which the branches of the tree appeared. This study of branching, called cladistics, has led to amazing discoveries. Your classificatory hierarchy—kingdoms, phyla, classes, orders, families, genera and species—are now some of the boughs and branches in the great tree of life.

In your writings, you speculated about how species can replicate themselves and how they might change, over time, into new species. In the middle of the 20th century, we learned that one of the molecules found in cells, deoxyribonucleic acid (called DNA for short), is the master blueprint. DNA can copy itself—it has two complementary strands, each of which can be reconstructed from the other. If there is an error during replication, the organism may change, possibly becoming new and different. DNA thus controls the relationship between parent and offspring, and explains the continuity (with some inevitable changes) that connects the branches of the tree of life. In fact, we can even use the changes in the DNA molecule to reconstruct the tree of life.

With 1.8 million species to analyze, each with many physical and molecular characteristics, it is now virtually impossible for humans to calculate how they all relate to each other. Instead, we utilize electrical machines known as computers to analyze these complex sets of information. For biologists, computers have become an integrated part of our experiments. Computers can compare the information encoded in a species' DNA, and its morphological and other characters, to hundreds of other species and thereby determine their evolutionary relationships.

Computers are also outstanding tools for organizing, analyzing and exchanging all kinds of knowledge. This year we began a decade-long project to use computers to assemble the Encyclopedia of Life (EOL), a collection of all that is known about the world's species. Each species will have its own 'page' on the EOL website.

The EOL will make it possible for the foundation of taxonomic knowledge to be available everywhere. EOL is partnering with the

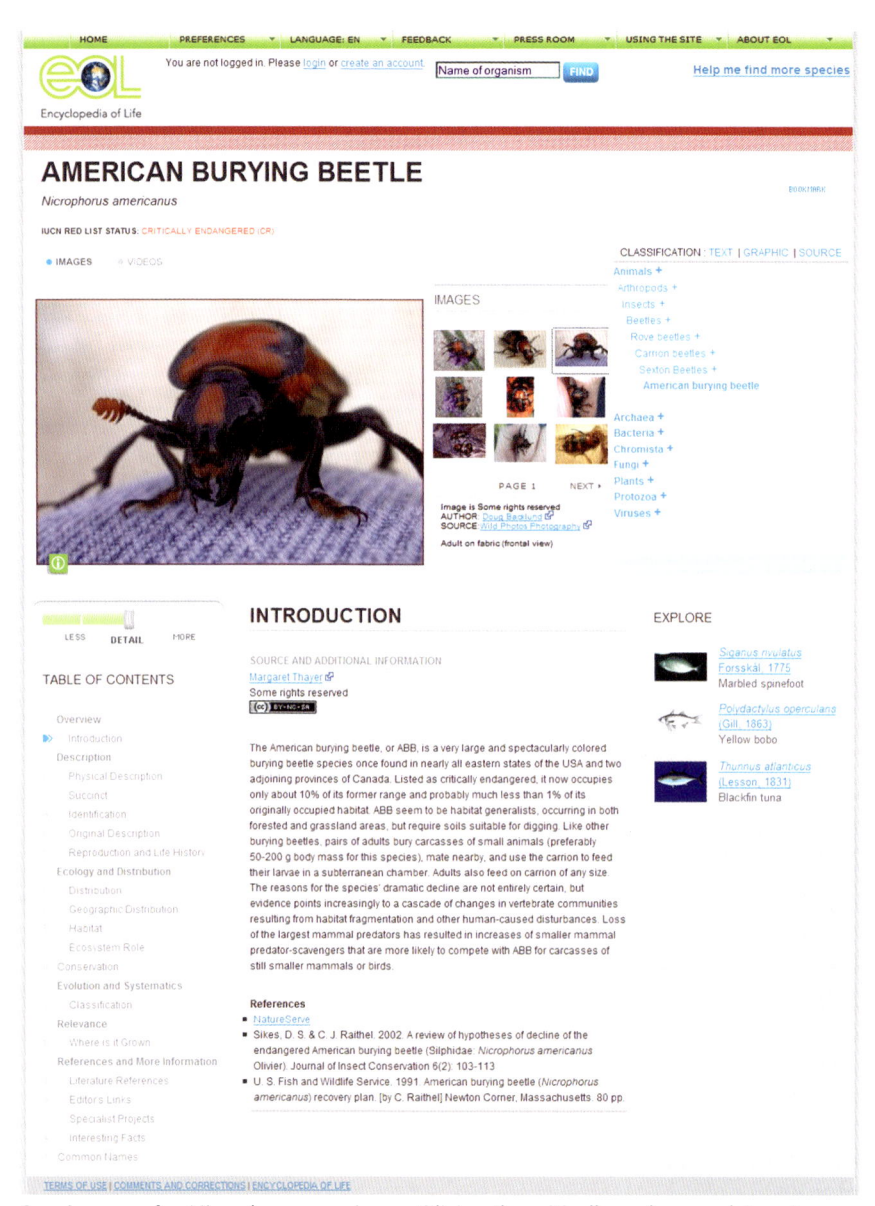

Species page for *Nicrophorus americanus* Olivier, the critically endangered American Burying Beetle (Silphidae). (Photo of *N. americanus* used by permission of Doug Backlund; introductory text by permission of Margaret Thayer.)

Biodiversity Heritage Library (BHL), a consortium of 10 of the world's largest natural history libraries, which are copying and making available the entire published literature on the world's biological diversity. A researcher will be able to see all the literature about a particular species, peruse the collected information about that species served by the EOL, and more quickly prepare taxonomic revisions and monographs. Today there are less than 10,000 taxonomists in the world, but this project seeks to provide them with electronic support of past and current research, while inspiring many more to join them.

The Encyclopedia of Life is not only a resource for research, but also a tool for education. Students will eventually be able to make observations on organisms in their schoolyard, connect their observations and questions to others with the same or similar organisms, and offer contributions to the refereed core of EOL. Schoolchildren will be able to compare their flora and fauna to organisms found on the other side of the world.

In essence, the EOL will be a 21st-century version of *Systema Naturae*, and everyone in the world will be able to use this resource. The beauty of the natural world, its wonder, and its ability to touch our lives, in good ways and horrid ones, will be available to all.

Today, we face many problems for mankind and the diversity of life as a whole. The growing population of humans has impacted the environment and its resources, and we are facing extinction of life forms at an accelerating rate. My hope is that this project will enable new types of questions to be asked about complex systems. Unlike microscopic queries of incredibly small things, the EOL will enable macroscopic questions to be asked about the Earth's species, and we will be able to visualize impacts on nature in a complete and global manner.

Sir, today you are known as one of the fathers to the study of life's forms. *Linnaean* principles are the means of conceptualizing order in the biological world, and I hope that you can visualize the impact you have had and will continue to have on humanity. In the *Systema Naturae* you invented the binomial system still in use for naming and classifying organisms and gathered together a wide range of knowledge about the world's plants and animals. We hope that the Encyclopedia of Life will continue in your great tradition to inform the world about the organisms that inhabit this planet with us, and will help to ensure that classification and systematics retain their fundamental place in biology.

Sincerely yours,

James L. Edwards

James L. Edwards
Executive Director
Encyclopedia of Life

Georg Dionysius Ehret (1708–1770) was a superb—some say the best of his era—botanical illustrator of German origin who worked with Linnaeus whilst he was with George Clifford in Hartekamp. The drawings in Linnaeus' 1737 *Hortus Cliffortianus* were done by Ehret, and Linnaeus showed him his "sexual system" of classification using the numbers of stamens in flowers. After working for Clifford for a short while Ehret moved to England, where he worked with Philip Miller of the Chelsea Physic Garden in London. Here he writes to Linnaeus from Chelsea in October of 1736.

He apologises for not sending the completed drawings of *Rauwolfia* and "Barba Jovis" [for *Hortus Cliffortianus*] sooner, and mentions two other drawings, one made through a microscope of "Tragia alia Scandens" and the other of which [above] which he thinks is a new genus with 10 stamens [*Erythrina crista-galli* L.]. He outlines his plans to travel to Jamaica with Miller, made with the assistance of others in English society [this trip never came to fruition]. He sends greetings to Linnaeus from Miller, relaying Miller's surprise not to have heard from him, and hopes the plants Miller gave to Linnaeus arrived safely in Holland.

Most learned Professor

I am only too aware that as a palaeontologist I work with mere fragments. At best I can only partially understand the animals and plants to which you have devoted your life. Without your work, the organisation and naming of the living world that we all adopt would be inchoate and inconsistent. You gave us a language so that we might speak to one another as fellow scientists. But with my poor fossils I fear that language will forever be partial, its vocabulary limited always by what can be preserved. Maybe it was an intuition that this was indeed the case that made you devote comparatively little time to studying the 'organic remains' in your native Swedish strata. You had already sought to embrace the entire living natural world: surely the story of what lay entombed in the rocks could await another narrator? The mere fact that you applied your binomials to fossils of trilobites or brachiopods surely shows that they would indeed one day find a place in your System, and that you acknowledged their biological origin in a way that was not always the case with your contemporaries. You were prescient as always, but even you could not anticipate how much, despite all its imperfections, the fossil record would reveal about the long history of life on our planet.

I wonder if systematisation of the strata may have followed upon your own organisation of Nature's products. Perhaps your example

Twisted layers of rock show the power of the Earth's movement

A fossil trilobite in the genus *Erbenochile* from the Devonian Period, 410 to 360 million years ago

in classification inspired Aldophe Théodore Brongniart and Georges Cuvier to organise the rocks of the Paris Basin into their natural successions? But maybe the march of canals across Great Britain was equally important, where the practical use of fossils to identify strata found ready appreciation from William Smith. I have just finished my term as president of the Geological Society of London, across the way from the Society that carries your own name. Every day I had the privilege to contemplate Smith's pioneering geological map that adorns our entrance hall. Indeed, maps—coloured maps—of strata became the geologists' guide to locating the treasures of the rocks; to unearth the story of life itself. Just as a handbook without your classification was an unwieldy thing, so geology without maps was lost in confusion.

If you could only have known what wonders would be revealed by excavation! Reptiles larger than elephants, preyed on by other reptiles with great sharp teeth! It is small surprise that at first they were described as 'monsters'—and portrayed in hellish visions. Strata yielded up extinct animals and plants in profusion and apparently without end. The animals I studied for more than thirty years, trilobites, now have more than 5,000 genera. I think you might have known them all under only one name: *Entomostracites*. Such is progress. But your binomial method continues to work even amidst this apparently endless profusion. Its joy is the apparent simplicity of your scheme, with its capacity to accommodate all the species, past and present, thanks to the flexibility of language. There have been attempts to supplant the Linnaean names, and doubtless there will be again, but any numerical system will fail because numbers are so unmemorable. Taxonomy is a tough job, and an unappreciated one in our philistine era, I regret to say, Professor. But one of its small pleasures is the invention of names; appropriate names, or entertaining names, or—dare I say it?—even beautiful names.

So fossils may not yield every morsel of information we may extract from the living organism, but what they do reveal is the unpredictable and remarkable course that life has taken. The track through time is tagged with the Linnaean names of animals and plants that are no longer with us on earth. When you were working on *Systema Naturae* it might have seemed that there would be no end to nature's bounty, nor that species could completely disappear. Now, we know only too well that mankind himself is driving his fellow creatures to the edge. The fossils tell us of times when there was mass death of many species, and that it took life millions of years to recover from these events. It could be that now we are witnessing an extinction of like magnitude fuelled by the inexorable growth in human numbers. Nowadays, zoos attempt to be monstrous conservation gardens rather like your systematic beds in Uppsala: a few examples of each kind to keep the species alive. I am sure you share with me the belief that species belong in the wild—that all of us and not just the two legged *Homo*, have and deserve a place on the earth.

If the fossil record provides a litany of species names devised according to your system—species that have already disappeared—then should it be a warning to us? Would you also not agree that it would be a tragedy for a species to go extinct now before it has even received the blessing of a Linnaean name? Like a child whose birth is unregistered and unrecorded, without a name there is no possibility of a biography. What you have given us, Professor, is nothing less than the materials with which we can write the biography of our planet. How I hope that these do not become the names on grave-stones in the cemetery of vanished species.

I wish the great organiser well.

Richard Fortey

VII-2008

Hello Carl!

Ok, one awkward moment down. Let me start by saying that I am a big fan of yours, though admittedly in a selective way that favors my own positions about taxonomy but mostly ignores many of your other accomplishments and occasional failings. Therefore I will not waste anyone's time with remembering the latter, but instead will let you in on some post-millennial developments in the field of taxonomy that you have prepared for us so brilliantly. It's a personal view on what your system of nomenclature means and where we are heading today.

As a matter of provision, I will take your taxonomic practice to best represent your more foundational positions about nomenclature. (I have some reason to think that this will not offend you too much; let's not revisit the theory of reproductive systems you used to create those 24 'plant' classes.) For example, I understand that a large number of your species descriptions were associated with particular designated specimens that were later reinterpreted as types. The 'type method' did not become a formal requisite until the mid-19th century, yet in some sense one could claim—as I prefer to—that you were an early practitioner. Similarly, when you classified fleas (which are secondarily wingless) closer to other winged insects than to ancestrally wingless insects, one could argue that you had an intuitive handle on such concepts as global parsimony ("what is the most plausible placement given all available character evidence?") and evolutionary reversal (acquisition, then loss of wings), even though there were no theories available at the time for you to properly accommodate these judgments. Undoubtedly some will say that this line of interpretation is giving you too much credit. I am sure there's a sophisticated reply to that charge. But just between us—giving you lots of credit is what big fans will do.

Fast forward to the 21st century. Near the turn of the millennium I began studying insect systematics for my doctoral degree in Ithaca, New York. Before that my understanding of nomenclature was basic, though I was fortunate to have had several years of Latin in high school[1]. As an incoming student at Cornell University I was instantly attracted to a particular school called phylogenetic systematics or

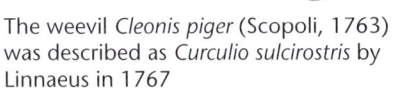

The weevil *Cleonis piger* (Scopoli, 1763) was described as *Curculio sulcirostris* by Linnaeus in 1767

cladistics. Cladists had come to identify themselves as a group some 25 years ago based on ideas that had more to do with the proper recognition of natural taxa than with rules of nomenclature. Beyond that they seemed to share a disposition (turned youth recruitment strategy) for talking tough, both to members and non-members, with varying motivations and results.

I soon realized that systematists are a dynamic and sometimes pugnacious bunch! There were a seemingly infinite number of historical and contemporary issues of contention to catch up on. How can we postulate natural groups? What are species? What *were* these authors smoking? I was sorting through the particular mix of posturing and solid argumentation that was going on in each case, attempting to identify my own mix. In retrospect my early attempts at defending principled stands about particular systematic methods were both fun and pathetic[2]. Naturally some positions changed through practice and experience. The initial cladistic chip on my shoulder morphed into a perspective where considerations of theoretical plausibility could win over a strict adherence to methodological purity.

There are two points I'd like you to take from this. First, my generation of systematists was trained by one that fought passionately for the recognition of their school. On many counts they've succeeded, and as a result cladistic methods are now widespread. The new generation is perhaps less combative, enjoying a certain status and confidence that weren't there for many years. This is not to say that all is rosy; new challenges pop up at every turn (as you will see). But I presume you would be pleased to see that we are still describing taxa in ways that are in line with the foundations you've laid, and that lately our field is reclaiming its vital status in biology. Second, although phylogenetic systematists radically opposed the taxonomic groupings of competing schools, at the time all schools accepted the system of Linnaean nomenclature. The applicability of your system to cladistics in particular is rooted in a shared use of hierarchical, tree-like summaries reflecting the character arrangement and relationships among perceived natural taxa. It would be hard to believe that Linnaean names survived so many episodes of bloodletting in systematics unless you had gotten something profoundly right about how we should describe and name natural groups.

Or perhaps not? Some 20 years ago a group of systematists launched a critique of numerous aspects of Linnaean nomenclature. These authors ultimately proposed an alternative under the heading of phylogenetic nomenclature. This letter will not allow me to do full justice to this alternative and its various refinements, nor will I have sufficient space to represent the diverse arguments, ranging from foundational to practical, that defend the Linnaean system against this newer proposal. Instead I will cherry-pick some bits and pieces from both sides to help set the stage for the main points that will follow.

From a certain perspective, our classifications have changed much over the past 250 years, and for too many groups there is no end in sight! Proponents of phylogenetic nomenclature have criticized many aspects of established nomenclatural practice, including a perceived lack of referential stability, the use of ranks (genus, species, etc.), the use of type species and specimens, and the use of phenotypic features to define species and higher taxa. Some criticisms are on theoretical grounds; viz. the alleged failure to differentiate between classes (which have essential properties) and individuals (which merely are parts of a whole), or the failure to define taxa in terms of phylogenetic ancestry. On the other hand, arguments against Linnaean ranks and other seemingly destabilizing rules for adjusting names are of a more practical nature.

So what's really going on? Surely, if we choose to abandon a working language, we had better be right about the nature of the problem, or risk throwing the baby out with the bathwater. Few systematists—or biologists, for that matter—think that the Linnaean system is perfectly suited for all communication needs. Advocates of any alternative system can thus exploit a certain lack of content with the *status quo*. But this does not free them from the task of finding out what shortcomings really matter, and how to overcome them.

It's a pretty safe bet that Linnaean nomenclature is *not* fundamentally incompatible with modern evolutionary or philosophical reasoning. It's relatively easy to see why. In defining taxa we traditionally make reference to particular phenotypic features; e.g., spiders have spinnerets. On the surface this may look as though we are thereby stipulating that taxa must have universal and immutable characteristics, which would amount to a non-evolutionary view. But perhaps we are just sloppy with our use of language. When we say "spiders are defined by their spinnerets" in a systematic treatment, what do we actually mean? Suppose a young lineage of spiders lost its spinnerets. We would still consider these taxa as spiders, adding in our minds a qualifying phrase to the original definition, such as "unless the spinnerets were lost in an event which—according to our most reliable phylogenetic estimates—occurred later on in a particular subgroup of spiders". What if spinnerets evolved elsewhere, say, in ants? Then we might clarify in this manner: "the term 'spinnerets' in spiders really refers to a series of (inferred, historical) transformations at the molecular level that are expressed phenotypically as silk-producing glands in a lineage whose sister group includes whip-scorpions, though not wasps". In the same manner, we could account for the loss of spinnerets at the molecular level by stipulating the origin of a secondary genetic transformation that inhibits the expression of spinnerets.

With the proper semantic modifications, the term 'spinnerets' can thus refer precisely to that series of historical evolutionary events that resulted in an ancestral spider species which subsequently underwent diversification into some 38,000 descendant species. Similarly, we can say that snakes are defined by their loss of legs, where 'loss of legs' ultimately refers to a secondary event at the molecular level that inhibits the expression of legs, thus making it a condition unique to a particular lineage of tetrapods and distinct from ancestrally legless vertebrates such as lampreys. The point is, while these referential refinements seem too cumbersome for everyday use, they clearly rely on modern concepts of evolutionary change and common ancestry. Indeed, many Linnaean definitions of taxa may benefit from such refinements[3]. There is no need, however, to regard feature-based definitions as incompatible with evolutionary thinking when the main villain is linguistic laziness.

What about abandoning Linnaean name definitions for philosophical reasons? As the argument goes, taxa are evolving and are therefore more like historical individuals than classes with stable properties. In truth, taxa are somewhere in the middle of these two categories. Their individual or class-like nature is more or less relevant depending on the kinds of inferences one intends to make in a particular context. The challenge for proponents of phylogenetic nomenclature is thus twofold. They must prove that (1) nearly all of the most critical contexts in which taxa are mentioned concern their nature as individuals; and (2) still more importantly, that applying the philosophical theory of classes versus individuals is more appropriate than using competing philosophical concepts that readily accommodate established practice, e.g., by relaxing the criterion of immutability in classes.

Let's look at an example of a property-free definition of passerine birds[4]: "Passeri are the most inclusive clade containing *Passer domesticus* and any extant species and including *Corvus monedula* but not *Tyrannus tyrannus*, *Pitta sordida*, *Furnarius rufus*, and *Thamnophilus doliatus*." Even though in principle this definition is independent of a tree diagram, we need to visualize some sort of tree to understand what "Passeri" means. Yet even then, what *can* we understand given such a phrase? Can we recognize a

The House Sparrow, *Passer domesticus* L. (Passeridae) from John Gould's *Birds of Europe* (1837)

member of the Passeri in a collection or in nature? Can we explain what Passeri are to colleagues, students, or to children? Can we associate with this name any prominent adaptations and relate these to other genetic, ecological, or behavioral information we have come to learn about passerine birds? If the answers are negative, then we have proposed a definition that is largely devoid of cognitive content and unable to support the kinds of inferences that biologists intend to make! At that point, when taxonomic names are no longer the primary vehicles for the causal phenomena we wish to learn and talk about—a situation that would seem unique in any science or language in general—it is time to question the philosophical wisdom that got us there.

Phylogenetic definitions such as the one above may hold stable across multiple succeeding taxonomic perspectives. Yet again, what are the pay-offs of such nomenclatural stability when the associated age, referential extension, and evolutionary properties of a name can change significantly from one phylogeny to the next? As a general guideline, nomenclatural rules should strive to bring out as much stability as possible in succeeding classifications, but should never be designed to obscure different hypotheses about phylogenetic relationships and thus weaken the name/taxon link over time. If a system of nomenclature is to remain stable in light of significantly diverging views about how nature is organized, then that system can no longer represent progress in systematics.

Let's move along. The consistent use of Linnaean ranks often leads to difficult real-life decisions. Seemingly minor adjustments in phylogenetic arrangement may require a series of inconvenient changes in nomenclature. Yet again, any practical shortcomings need to be weighed against the immense inferential benefits of forming a match between temporally succeeding evolutionary events and hierarchically nested taxa and names. For example, by knowing that a spider is a member of the Salticidae, one can infer more

The Jackdaw, *Corvus monedula* L. (Corvidae) from John Gould's *Birds of Europe* (1837)

than 158 million three-taxon statements of mutually exclusive groupings among spiders alone[5]. As inductively working animals, we simply like these kinds of cognitive shortcuts too much to abandon them for a language that requires infinitely more acts of unconnected memorization.

You see, Carl, your system has come a long way. I venture to say that this was not an accident, or a product mainly of convention. Your proposal to embed hierarchical information into Linnaean names, to root these names in both physical specimens (ostension) and in perceived phylogenetically relevant properties (intension), is fundamentally sound. It is sound and successful not because it follows the tenets of an appealing philosophy, but because it is particularly well suited for human learning about nature and for accommodating the kinds of inferences that are of primary interest to biologists. Accommodation is the capacity of a linguistic tradition to align itself with the causal structure of the world and thus enable efficient communication about relevant natural phenomena. A successful language must excel at accommodation, a quality that emerges gradually over time.

Linnaean names have an impressive track record of facilitating scientific progress across all biological disciplines. This list certainly includes evolution, as it turns out that evolutionary biologists focus much of their research on the distribution, temporal sequence of transformation, and biological significance of *organismal properties*! To uphold their practice over time these biologists must rely on the responsiveness of taxonomic names to newer and more accurate hypotheses about the phylogeny of a group of organisms and their characteristics. In other words, the process of 'voting' for or against a system occurs in no small measure outside of the context of systematics and philosophy. Linnaean names had to pass many external and independent trials of accommodation in order to establish and maintain their current standing.

Still, all is not well today. Clearly, we should have a more thoroughgoing philosophical account explaining why and to what extent Linnaean names and their semantic components succeed at accommodation. That account is limited, however, by how much we understand about successful reference in general. We know for instance that Linnaean names are of a hybrid nature that tends to yield more than the sum of its parts. Accepted naming practice involves an event of baptism, e.g., the type designation for a perceived taxon. Such an event can trigger a causal chain of speakers who may understand each other even though everybody misjudges what characteristics pick out the taxon among its relatives (= reference in spite of misdescription). Similarly, the property-based description of a taxon may be incomplete or imprecise but performs sufficiently well in a number of relevant situations (= partial reference). And so the acts of pointing and characterizing jointly refine the meaning of taxonomic names.

Unfortunately, that's not all. My limited knowledge of philosophy of language has taught me that humans have a knack for understanding each other even when prominent philosophical theories say they shouldn't! Success, partial success, and failure in communication are profoundly situational. Each outcome depends in part on whether one speaker makes the right assumptions about the other speaker's training background and present usage of a taxonomic name. To make matters worse, we rarely provide more contextual information than we assume is needed, preferring instead to adopt certain usages of terms until a misunderstanding is apparent. This ubiquitous habit of ours presents a challenge for philosophers, who must account for contextuality and variable underlying assumptions in order to explain why reference can work in some cases though not in others. It also constitutes a real-life problem for systematics, especially in taxonomic groups with a history of significant rearrangements.

Which leads me to my main point. The issues we're tackling today with Linnaean names are not really rooted in the naming process *per se*. I think that proponents of phylogenetic nomenclature correctly sensed that there was a problem, but got the diagnosis mostly wrong. The real issues arise through a combination of (1) how the naming of taxa is legally regulated (through the Codes, etc.), (2) how these rules are implemented and supplemented with additional information, and (3) how these two processes interact over time. Many users who are unsatisfied with 'the system' primarily feel that there is a lot of baggage in taxonomy. It's difficult to impossible to sort through that baggage, leading to linguistic imprecision or even paralysis in certain taxonomic groups. These users have a point, though the problem is more likely rooted in a history of inadequate systematic inferences and poor linguistic implementation than nomenclatural rules.

Let's be honest, Carl, by the time your later editions of the *Systema Naturae* were published, you thought you had a solid handle on a large chunk of nature. Well, allow me to say that in the case of weevil species, you finished short of the mark by a factor of more than 2,000! So with less than 1/2000 or 0.0005% of the total species-level diversity on hand, we had a system in use. It seems preposterous, in a way, but then not much has changed. Even today, we have a tendency to advertise systematic works as definitive (see 'the tree of life' or other 'synthesis' projects) when in reality we're still stepping in the dark. That's how science works in a competitive world where the first and the loudest reap the most benefits for their program. The Linnaean system is hardly at fault here. And in case we're closer to the light in terms of adequate taxon sampling, we might still have inadequate methods for phylogenetic inference. Remember, the most powerful concepts and tools for generating phylogenetic trees are less than four decades old. Many groups have not yet been subjected to these methods. Then there's poor taxonomic work or judgment in spite of good sampling and methodology. It happens—with or without using Linnaean names.

In short, much nomenclatural baggage over the years is due primarily to work that looked sufficiently decent at the time but was ultimately too far off the mark to remain in use today. We'd be hard pressed to find any practical remedy for this phenomenon. New rules for naming won't change the way we (over)confidently regard and sell our products. There's also no point in holding off too long with a new system. We simply can't get from nearly 100 recognized weevil species (*Systema Naturae* in 1758) to 62,000 species (today) to 220,000 species (recently estimated) unless we split up the task and recognize the utility of small, incremental gains in phylogenetic knowledge. It's a sensible risk/reward strategy—something wins out over nothing—but naturally we pay a price for the resulting taxonomic and nomenclatural baggage.

So while we shouldn't abandon the ground rules and can't seem to escape the costly strategy of gradual increments, I think there's plenty of room for *strengthening the semantic ties* among multiple succeeding classifications. If systematists can't guarantee stability in meaning then we should at least offer more transparency. As experts we can make explicit our underlying assumptions, new insights, and differences with former systems on a much more regular basis. I will give you an example. Two years ago I published a phylogenetic revision of a weevil tribe (Derelomini Lacordaire) that now includes some 40 genera and 270 species. In that work I transferred 11 genera into the tribe that previously were placed in 4 other tribes. I also transferred 6 genera that had been previously part of the tribe into 4 other tribes. In all, 17 generic rearrangements were made, and 7 tribes were affected including the tribe I revised. Now, when I say "previously", I mean the classification presented in a particular weevil catalogue that was published less than a decade before. That much was made clear in my revision. I also attempted to highlight the proposed synapomorphies for the tribe, including reversals in select nested lineages, and avoided presenting them along with phylogenetically less relevant diagnostic features. So I was aiming for some level of semantic precision and transparency.

But then I knew so much more that remained unmentioned. For example, I had studied the taxonomic history of virtually every genus and species in the tribe, and could have rigorously traced its taxonomic placement in any major revision from 1798 to 2006. I also understood the relevant property-based definitions of the tribe published from 1866 onwards, and could have specified the extent to which they overlap among each other and with my perspective. This sort of information would help tremendously in terms of reconciling past and present taxonomic perspectives, thereby reducing the semantic

A tray of weevil specimens from the Linnaean collections at the Linnean Society of London

inconsistency and ambiguity that has characterized the taxonomic history of the tribe. Alas, this is almost never done. Furthermore, I should have presented a more formalized taxonomic update of the six non-focal tribes that underwent rearrangements in my study, by presenting side by side the constituent genera of each tribe according to the previous and my revised perspective. For further semantic disambiguation I could have added statements such as "tribe X [previous perspective] *corresponds* to tribe X [present perspective], *minus* genus Y, and *plus* genus Z". There are ways to express this so that virtually any user or specialized computer software can infer the full set of similarities and differences between the two perspectives.

We're at a juncture in systematics when more precise phylogenetic estimates are published at an increasing rate. There is a concomitant trend to archive the results in networked repositories intended to serve as the primary 'hubs' for systematic information[6]. Both the systematic and the computer science community seem to have bought into this vision. However it is likely that each community underestimates just how much we need to adjust our linguistic habits in order to achieve long-term integration of systematic products. Computer scientists use a formal language (description logic) to build highly structured networks (ontologies) that may include classes, instances, parts, properties, relationships, and other components and qualifiers. Once the structure is in place then powerful algorithms can 'reason' about the constituent elements, connect them to other ontologies created for related subject areas, and so on.

As computer scientists learn about systematics they must initially see a strong match between an ontology and a published taxonomy. However, as we've seen, a classification is never entirely comprehensible in isolation, and instead represents a complex mosaic of previous and new elements with implicit identities and relationships to each other. Too often such expert-made classifications are only comprehensible to other expert speakers, i.e., persons who share an intimate understanding of the contextuality of the new system and are thus able to make explicit the implicit semantic links to previous systems. So I predict that early generations of ontologies for systematics will either permit a very limited number of automated inferences, or will rely very heavily on expert input in order to reason among multiple succeeding classifications. In short, computer ontologies and systematic practice are not yet ready for each other.

Why have systematists relied so much on painstakingly acquired, implicit assumptions about the taxonomic history of particular groups when presenting their new classifications? I believe the reason is neither some form of elitism ("take that, users!") nor a lack of self-esteem ("who wants to read about all these subtle similarities and differences?"). More likely, it's simply human habit—we make things just as explicit as we think is needed at the moment—paired with the similarly human notion that the latest perspective is really the one that's going to last for a long time, in spite of all historical evidence to the contrary. And so we pass the burden of full semantic resolution, both looking backward and forward, on to future specialists.

Let me try to sum up. The semantic problems we are confronting today in systematics are the result of a complex and long-winded interaction between accepted Linnaean practice for naming taxa and the particularities of taxonomic work that have piled up over centuries. The solution lies primarily outside of the Linnaean system of nomenclature as implemented today. The latter has served us mightily in accommodating inferences about the perceived properties of taxa. It generally mandates a hybrid model of intensional and ostensive definitions which, if properly interpreted, are compatible with evolutionary thinking as well as modern theories in the philosophy of language. Adhering to Linnaean ranks has allowed us to learn and communicate about nature in a way that suits our mental capacities as well as the causal structure of the natural world. We can only thank you for that, Carl.

However, the Linnaean system is *not* capable of capturing the *entirety* of semantic adjustments that occur when a previous classification is revised in light of new evidence. In fact it was purposely designed to respond to some kinds of taxonomic rearrangements but not others[7]. Instead of abandoning the Linnaean system, this observation should lead us to express more clearly and more consistently what we mean when presenting a new classification. We must invent ways to semantically map each component of the new system to the corresponding component of a relevant predecessor, stating all intensional and ostensive similarities, differences, and apparent ambiguities. In doing so, we will reduce the sense of baggage in systematics and make progress towards a full semantic integration of the taxonomic process (not just individual snapshots) via ontology-driven services. At the human level, this requires that we routinely acknowledge the ephemerality of our latest insights, spend more time comparing our perspective to a previous one that we no longer think holds true, and generally pay more attention to the context in which we use taxonomic names. Efforts to achieve this are presently underway and are summarized under the term 'taxonomic concept approach'[8]. If we *supplement* the Linnaean system with these conventions, there will be more linguistic transparency and less mistaken urgency to purge the idiosyncrasies of the past or legislate a wrong consensus. So let's start to take our semantic supplements; they're all that's really needed to successfully use Linnaean names for the next 250 years!

In deep admiration,

[signature: Nico Franz]

Nico Franz

Notes:

¹ I suppose that most people take for granted the balance of beauty and information of your binomials. I have never heard the phrases "Linnaeus awkwardly named . . ." or "Linnaeus inexplicably named . . ." Part of the Linnaean success story is the cognitive and esthetic appeal of the actual names you've created.

² Unfortunately, none of this relieved me from the difficult task of working with weevils. You may be surprised that we now recognize 5,800 genera and 62,000 species of these little snouted beetles, up from 2 genera and 94 species described by you in the *Systema Naturae*.

³ There are limits to this practice. As the number of convergent characters and evolutionary reversals increases within a lineage, the referential ambiguity will increase as well. Property-free definitions are perhaps the best option for some lineages such as bacteria, where high rates of transformation and horizontal inheritance obscure phylogeny.

⁴ Adopted and slightly simplified from Sereno, P.C. 2005. Systematic Biology 54: 595–619.

⁵ See Platnick, N.I. 2001. http://www.systass.org/archive/events-archive/2001/platnick.pdf

⁶ Some authors, typically working outside of the systematics community, have suggested that systematists should stipulate 'consensus classifications'. We'll grant that wish right after there's a unified and stable view on pressing topics in ecology or other biological disciplines.

⁷ Some measure of semantic ambiguity is desirable in a language that is reflective of small and often short-lived increments in knowledge.

⁸ For a not-very-humble reference see Franz *et al.* 2008. In Wheeler, Q.D. (Ed.): *The New Taxonomy*, Systematics Association Special Volume Series 74. Taylor & Francis, Boca Raton, FL; pp. 63–86.

Dear Dr Linnaeus

David Lodge, in one of his comic campus novels, has a character write a thesis on the influence of T.S. Eliot on William Shakespeare (I am assuming you are kept supplied with contemporary reading matter in your taxonomists' Valhalla and that these names mean something to you). His half-serious point was that we read *Hamlet* through the prism of the *Four Quartets*, and so it is impossible for us to know a pure Shakespeare unpolluted by his literary descendents. And I suspect the same is true of you; only someone ignorant of the taxonomic movement you founded could really understand the historical Linnaeus. You, like Shakespeare, and nearer home Isaac Newton and Charles Darwin, have become iconic, a palimpsest upon which we sketch our current preoccupations, hoping that a little of the authority of a scientific Immortal rubs off. I suspect some of the other letters you receive will be less intellectually lazy, will seek to delve beneath the two-fold image (the brand you skilfully created for yourself while alive and its subsequent burnishing by your disciples). But I lack the scholarship—never had the Latin. Instead I am shamelessly going to enlist your support for a couple of my hobby horses—I nervously await the envelope stamped "Return to Sender".

So cast your mind back to those brief Swedish summers of your youth, when you began to navigate your way through the (admittedly rather modest) Scandinavian floral diversity. You clearly saw the need to try and impose some order on the haphazard system of naming plants, on the capricious classification that grouped species together by their usefulness to man, and to provide a resource that would enable not only your fellow botanists but a broader community of people to identify plants and study them further. You were further stirred by the reports of the much greater diversity of plants outside your native land, some of which you saw alive on your European travels, others as dried specimens in the herbaria you visited. And there were even more animals than plants.

What you and your contemporaries were facing was the first bioinformatics crisis. The amount of information becoming available about plants and animals far exceeded the capacity of your current storage systems—herbals and bestiaries. The prevailing climate of the Enlightenment gave you a license to innovate, not forever to be in thrall to Aristotle and Pliny. Your solution, as you yourself pointed out many times when alive, was brilliant in its scope and, yes—again as you predicted, has stood the test of time. Suitably modified by your intellectual descendents we have been able to name over a million species of plants and animals—a triumph of human endeavour. While your

Protea lepidocarpodendron L. (Proteaceae) from Linnaeus' herbarium

detailed classification has not survived (though I personally mourn the disassociation of the Linnaean *Vermes*) you showed what needed to be done, and Charles Darwin provided the tools.

Lumbricus terrestris (L.) (Annelida), the earthworm, placed by Linnaeus in his animal category "Vermes"

In addition to nomenclature and classification, in *Systema Naturae* you provided a one-stop-shop for the biodiversity scientists of your day. You wanted not only experts to be able to understand your classification but anyone who had a need or a desire to identify plants and animals. You understood the relevance of stakeholders and user-groups before these words were invented, of the importance of making your research accessible and relevant to the broadest community of professional and non-professional biologists, of building up a constituency of supporters all banging the drum for more taxonomy. And the network of students you sent off to survey biodiversity around the world (at least those that survived—did you ever fill out a health and safety assessment form?) attest to your skills as a science administrator and leader, a true visionary.

It is in making taxonomy accessible to the greatest number of people that I think we have most failed to live up to the example you set. The reason for this is completely understandable, and until recently insoluble: there are hugely more species of plants and animals than you or your contemporaries ever imagined. Walking through those Swedish meadows did you ever think there were over 30,000 Swedish insect species, or sorting through herbarium sheets in Uppsala that there were about 350,000 species of flowering plants on earth? You had not been dead many decades before the notion of a new edition of *Systema Naturae* became unthinkable—it would have to have been massive, and beyond the capacity of any individual or institution. The enormity of the taxonomic task exerted a strong fissiparous force—the subject became fragmented and specialised, each subfield with its own specialist jargon and complex terminology. Taxonomy expanded hugely, and different workers often disagreed on how to name and classify different groups. Your successors devised complex codes and systems of rules to keep the Linnaean show on the road, with great success, but at a cost. The taxonomy of any major group today does not exist in one place in a modern *Systema*; instead it is in a distributed form—an ill-defined integral of all the papers published on the group, sometimes in many languages, often in obscure places. Being an expert on a group is as much having the skills to navigate this information space as it is to know the organisms concerned. Taxonomic research is difficult to use, and hence today hard to fund. Your descendents find it harder to send out students to survey the world's biodiversity, even if more survive!

So if you returned today, took time off from compiling that flora of the Elysian Fields, I think you would berate us for allowing taxonomy to become so disconnected from its users. You would point to the huge demand for information about life on earth, to the need for a modern *Systema Naturae*. Brazenly, I think you would be a supporter of unitary web taxonomies, of the idea that the taxonomy of a group of animals or plants should exist in one place, on the web, accessible to all. Instead of users requiring specialist knowledge and specialist libraries they would access the complete taxonomic corpus as long as they could connect to the web. You would argue that the website should be a platform for taxonomic research, for different taxonomists to put forward different hypotheses to be decided by new data (I know how much you enjoy chatting about this with Willi Hennig). Yet you would also encourage the designation of consensus taxonomies (which is what *Systema Naturae* was) so that non-specialist users need not navigate through the complexities of multiple hypotheses. The consensus taxonomies would be provisional and evolve over time as new data become available, and for some species groups may just say that no consensus is possible without further research. With your keen science-administrator eye you would spot the two major hurdles to making unitary taxonomies a reality: the need to persuade taxonomists that it is evolutionary not revolutionary and good for the subject, and the need to raise substantial funds to leap the digitisation hurdle. We could do with your presence at the conferences and fundraisers!

Revenants traditionally have two roles: to advise the living and to issue warnings. I also see you in more sepulchral garb stalking the land, tolling a bell with clanking chain: "Change your ways before it is too late!" you cry, "Don't forsake me!" You would go on to warn us about the future of taxonomy in an age when massive-throughput sequencing is very cheap. Already groups are experimentally sequencing whole ecosystems, and we are perhaps a decade, probably less, from this becoming commonplace. The inevitability of this is certain; what we don't know is its implications for taxonomy. But today, for the first time, it is possible to conceive of a means of classifying life on earth that does not involve Linnaean systematics. Species could simply be named as co-ordinates in sequence space or numbered twigs on phylogenetic trees: it has already largely happened in the prokaryotes. Does it matter if this transpires—well, probably not for taxa where Linnaean taxonomy is still rudimentary—prokaryotes, some meiofauna for example? But for the majority of plants and animals it would be tragedy if two disconnected systems developed—taxonomy as we practice it today, and purely sequence-based taxonomy. There is not the money for two systems and though no one will issue an edict closing down Linnaean taxonomy, death-by-a-thousand-cuts is just as fatal. I think you would exhort us to modernise taxonomy, making it more accessible, more relevant, and more efficient, so that the sequencing tsunami strengthens the field rather than washing it away. In another 250 years I want your taxonomic descendents still to be writing to you, not Craig Venter!

Yours faithfully

Charles Godfray
Charles Godfray

Carl Linnaeus

Goethe (Ebach)

Weimar, August 28th, 1819

My dear Carolus,

On the occasion of my birthday I partake in a little ceremony involving my favorite Harzkristall cut glass and two bottles of Rhine wine. Carefully placed at either end of my study (one towards the library and the other on the table overlooking my garden), I toast the achievements of naturalists, poets and artists. My first, dedicated to Nature and taken by the window, is followed by a toast to you. Standing there in front of my library I notice the dust and loose leaves of scribblings that have settled on rows and rows of my monographs and folios. One stands out—*Systema Naturae*—decades of thumbing through its pages has left it stained and grubby. I can picture each page perfectly, your careful observations, my notes in smudged lead.

As I stand here looking into my garden I think of the *system*, how I arranged my beds group by group and how it struck me that Nature herself orders life so differently, so organically somehow. I stand and gaze helplessly and watch Nature order herself—sometimes I pick out two plants of the same genus, maybe three, and wonder what the *urpflanze* or the archetypal plant would look like. I can still feel the excitement as I wandered in the gardens of Padua, searching, seeking, sensing the *urpflanze*. Oh how naïve I was! I smile at that simplistic naïvety as I behold the shambles that has become my beloved garden. Oh, if only I could find that single specimen or *urpflanze*. What an idea! As [Friedrich] Schiller himself had reminded me, the *urpflanze* is all idea—a reality within our *Anschauung* (intuitive perception). I wonder if you too saw that my dear Carolus, the *urpflanze* coming into being, a real natural entity like a mammal that is all mammals we have ever seen and experienced? O *Systema Naturae*, so close to finding order, but so far from Nature herself!

Order is not God given, not whole. Parts, be it stamen or other organs, added together do not make a whole like bricks in a house or coins in a mint. The whole is there already to be discovered through *Anschauung* of the parts and their relationships. Perhaps science must go through four epochs of science in order to understand the whole, the *urpflanze*, or the *urphenomenon*?

Four epochs of science:

childlike,
poetic, superstitious,
empirical,
searching, curious;
dogmatic,
didactic, pedantic;
ideal,
methodical, mystic

When we venture from the childlike to the empirical, we venture from our own subjectivity into the realm of Nature. There we can find the *urpflanze* and a way toward a natural classification. It is the oppressively dogmatic that I hope I shall never see. A classification that leads straight back to Immanuel Kant—that great Sage of Königsberg—and back to a 'natural' teleology. Will natural classifications survive such a dogmatic approach, one that attempts to overthrow one artificial system with another? I hope that the fourth and final epoch will end that madness and restore a natural classification and terminate that

Sea Serpent as depicted by Conrad Gesner in Volume 4 of his *Historia Animalium* (1551–1558)

desire for Natural systems—a contradiction! Nature does not have a system; we can never find its center nor its end. Our observation of Nature is limitless and our goal is to experience the whole, the *urphenomenon*. Dare we shy away from our own senses for they do not deceive; only our judgment deceives.

I feel for those trapped, like wasps in honey, in their own argot of useless terms. What is a goblin, ghost or Will-o'-the-wisp but names? I dare the superstitious to show me a banshee or an ogre; I dare the naturalist to show me a unicorn, sea serpent or griffin. These are imaginary creatures, simple fiction—so too are systems. Why do naturalists attempt to kill the phenomenon with a word or worse still, create a term in place of a phenomenon where one does not exist? The answer lies in our distrust of our own senses.

Our own appreciation of colour, its effect on our senses and in the presence of other colours, whether they are solid or light, is degraded with Isaac Newton's mechanics. No longer is the colour blue something that directly affects our senses, something that cannot be measured without them; instead it becomes a clutter of numbers and fractions. If we reduce all qualities, such as those of the flower, to mathematical equations we are left with nothing but a few fractions—meaningless to a comparative anatomist. If comparative biologists wish to seek comfort within the whole organism, they must learn to discover the whole in the smallest part. Only our own intuition and sense perception will allow us to do that empirically.

You have appreciated Nature, Carolus! She understands no jesting; she is severe, true and always right. The errors and faults are ours alone. Nature despises those who are unable to appreciate her. She only reveals herself to the able and the true and to you.

J.W. Goethe

Notes:

Malte C. Ebach for Johann Wolfgang Goethe (1749–1832). Translations taken from: Goethe, J.W. von. (1995). *Scientific studies*. (D. Miller, Ed. & Trans.). New Jersey: Princeton University Press and; Naydler, J. (1996). *Goethe on Science: A Selection of Goethe's Writings*. Edinburgh: Floris Books.

Carl Linnaeus

Antoine Gouan (1733–1801) was a French botanist and natural historian who lived and worked in Montpellier. He wrote an early catalogue of the plants of the Montpellier Botanic Garden, the first work in France to use the binomial system. He was one of Linnaeus' more frequent and cherished correspondents. They exchanged seeds and bulbs, and Gouan often enclosed fragments of plants in his letters for Linnaeus' examination. He used long polynomials when describing new plants, despite being one of the first to take up the binomial system after its introduction in *Species Plantarum* (1753). In this letter, written "To the most Distinguished Von Linné" from Montpellier in October of 1770 he sends detailed descriptions of some umbelliferous plants from the Pyrenees.

86 *Letters* to *Linnaeus*

After Linnaeus' death in 1778, Antoine Gouan continued to correspond with Carl Linnaeus filius (the son of Linnaeus), and the seed and bulb exchange seemed not to cease. In his letters both the Linnaeus and to his son, Gouan included not only plant fragments, but also occasionally exquisite drawings of insects and very occasionally actual animal specimens (see fish above)! The date when these enclosures were sent is not known.

Viro Sapientissimo Professori Carolo Linneo Felici Salutem Plurimam Dicit Andreas Hamiltonius

I know from speaking with my colleagues in several disciplines that you've been receiving a lot of correspondence lately. As an Assistant Professor with almost my whole career ahead of me, I find it appalling that you have so many calls on your time so long after your retirement, but I hope you will indulge one more.

The recent tricentennial of your birth had many of us—historians, philosophers, and taxonomists—discussing your legacy, as well as the conceptual structure of taxonomy generally. The main purpose of my letter is to tell you about your impact on historical and conceptual debates. You know, I'm sure, that you've had an incalculable influence on how biology is practiced, but the importance of your work for understanding science generally has, sadly, not been much of a topic of conversation lately.

There are reasons for this silence that don't have anything to do with you. It strikes many of us in the life sciences as odd and strange, but sometime near the end of the 18th century it was decided by commentators of several stripes who might collectively be called natural philosophers that Newtonian mechanics should be the paradigm of good science. Your younger contemporary Pierre-Simon, Marquis de Laplace was among them, as was your more direct contemporary Pierre-Louis Moreau de Maupertuis.

They and very many others who were of a mathematical turn of mind were impressed by the notion of natural laws and of explanations given in terms of what the mathematics says *must* occur given some initial conditions. Why should a comet appear in 1758—just as Isaac Newton had predicted? Because there is a universal attractive force. How do we know that there is a universal attractive force? Well, assuming that there is one allows us to make predictions about things like comets, the motions of the planets, and the tides. Newton's mathematics—his models—explain much if

The orbit of the moon influences the tides on Earth

only we will accept the assumptions that Newton himself was not willing to feign.

Some, Maupertius among them, realized that this form of reasoning had important problems. His particular solution was to take two approaches: as a mathematician he was pleased to get predictions that could be tested using Newton's framework, but as a scientist he had serious doubts about "whether Attraction accords with or is contrary to sound Philosophy." Action at a distance, indeed! Somehow this concern has mostly been lost, and we were left with the notion that what scientists should do is find laws of nature that are universal and expressible in simple mathematics. We are also left with much less concern about whether models limn the nature of reality or whether they just provide a means to make good predictions. This was a central question asked by Newton and Maupertius, and for that matter by Andreas Osiander when he wrote in the preface to Nicolaus Copernicus' *De Revolutionibus* in 1543 that "these hypotheses need not be true nor even likely. On the contrary, if they provide a calculus that is consistent with the observations, that alone is enough." *Hypotheses non fingo.*

Your work reveals you to be quite at home in a complex world, so this reasoning will probably seem as strange to you as it does to me, but the situation gets worse. In the early 20th century, professional philosophers who didn't know any biology formulated accounts of how science works, or should work, that codified a need for natural laws and for explanations in terms of predictions based partly on these laws. At the same time, historians become seriously concerned with scientific revolutions and their causes as well as their implications. Both groups eventually fixed on theories as the relevant conceptual structure to analyze. They wrote libraries full of arguments about what scientific theories are and how and why they change, but most failed to notice that not all revolutions have a basis in the kind of theoretical structure they were studying.

You will have worked out by now why I am telling you this story about historians and philosophers: having fixed on something that used to be peculiar to mechanics as indicative of what all science is and how it ought to work, they articulated what is now a standard list of the revolutions in science since Copernicus. Your name is not on the list. Again, this has nothing to do with you, but rather with a focus on the physical sciences, quantitative models taken to be laws of nature, and attachments to a certain sort of scientific theory. It may help you to know that Dmitri Mendeleev did not make the list either, and for the same reasons.

This may be changing. I have been encouraging students who are interested in the history of science to write about scientific revolutions that are system or catalogue based. Though none of them have yet taken the bait, you may still get to be a revolutionary. Everyone agrees that your system of taxonomy and Mendeleev's periodic table had revolution-sized impacts, but there is lingering misunderstanding about what these ways of organizing the world come to. The general sense seems to be that you and Mendeleev found ways to sort what was already known into useful categories rather than making original contributions to human knowledge. I'm sure this comes as a surprise to you, as this is not what you thought you were doing. (I know this because I have been reading your mail.) Mendeleev didn't think so either, and he captured his (and your) project quite nicely:

> The mere accumulation of facts, even an extremely extensive collection . . . does not constitute scientific method; it provides neither a direction for further discoveries nor does it even deserve the name of science in the higher sense of that word. The cathedral of science requires not only material, but a design, harmony . . . a design . . . for the harmonic composition of parts and to indicate the pathway, by which the most fruitful new material might be generated.

We may yet realize that good taxonomy, biological or otherwise, is not only about sorting things out, but also about providing the very conceptual structure by which we might approach the world. Mendeleev's periodic table is not only synthetic, but also predictive, as Paul-Emile Lecoq de Boisbaudran's discovery of gallium in 1875 showed.

Your taxonomy, coupled with what we now know about patterns of distribution of organismal characters and evolutionary processes, regularly makes good predictions as well. Given two closely related species we can know quite a lot about what their common ancestor was like, even if the ancestor species is no longer extant. We have built a spectacularly well informed—though far from complete—tree of life partly on this basis. We can also make good practical inferences. Paclitaxel, a plant alkaloid isolated from Pacific yew trees (you would call them *Taxus brevifolia*), has been used to treat breast and ovarian cancers for ten or fifteen years. A concern for conserving these trees had us looking for new sources of paclitaxel not long ago because *Taxus brevifolia* was the only known source. Taxonomic knowledge constrained the search, and paclitaxel was found in other species of the same genus.

As you might rightly point out, understanding the biological world in terms of species within genera within higher taxa was your innovation. While the scheme in your *Systema Naturae* may look to the uninitiated as a kind of Dewey decimal system for plants and animals, some realize that your approach is *conceptual*, and that just as with Copernicus, Newton, and Mendeleev, it makes testable claims about what the world is like while providing a system of organization for old and new discoveries. Your taxonomic system makes claims about what kinds of biological objects there are and how these objects are related to one another. This is precisely what Mendeleev did for elements and precisely why both of your projects precipitated scientific revolutions in the fullest sense of the phrase. They changed the way we understand what there is and what systematic relations hold between objects in the world.

As with Copernicus and Newton we can ask whether your system gets it right—whether the structure of your taxonomy captures and expresses the structure of the biological world. Here lies the crucial difference in what you did and what some take you to have done: imagine asking whether Melvil Dewey's organization of knowledge into ten main classes, a hundred divisions, and a thousand sections correctly describes human knowledge. This isn't even a well-formed question, because knowledge doesn't really have joints and because Dewey's system is a convention for organizing that does not make claims about branches of human knowledge and how they relate to one another.

There is one more thread about revolutions that is worth mentioning. Most Americans were taught in secondary school that there is a single scientific method, and were also provided with an embarrassingly impoverished idea of how scientists carry out their investigations. This may have something to do with the continuing but silly worry that taxonomy is not an experimental science, and therefore not science enough. You and I might point out that Newton didn't do any data collection (much less experimentation) to arrive at his universal theory of gravitation, and that very many good sciences are necessarily limited in what they can do by way of direct experimentation. Cosmology comes to mind, as does paleontology. From the assumption that all good scientific investigations must be experimental, it follows that these are all second-rate sciences. Good reasons for this assumption, however, escape me. Is there really reason to think that lab-bench experiments of a certain kind are the only (or even the best) way to test every kind of

Portrait of Isaac Newton whose *Philosophiae Naturalis Principia Mathematica* (1687) presented his theory of universal gravitation

hypothesis? The history of science—both recent and not-so-recent—strongly suggests that the answer is 'no'.

Thank you for taking the time to read a bit about what historians and philosophers have been up to over the last hundred years or so. I'm hoping that we can do a better job of understanding what you and other system builders contributed to science and thereby do a better job of understanding how science works. You should have convinced us that there is more to science than theories of a certain very narrow sort, and that sometimes the important conceptual breakthrough is taxonomic in nature.

Vale, Magister. Di te incolumem custodiant.

Andrew Hamilton

Andrew Hamilton

Carl Linnæus

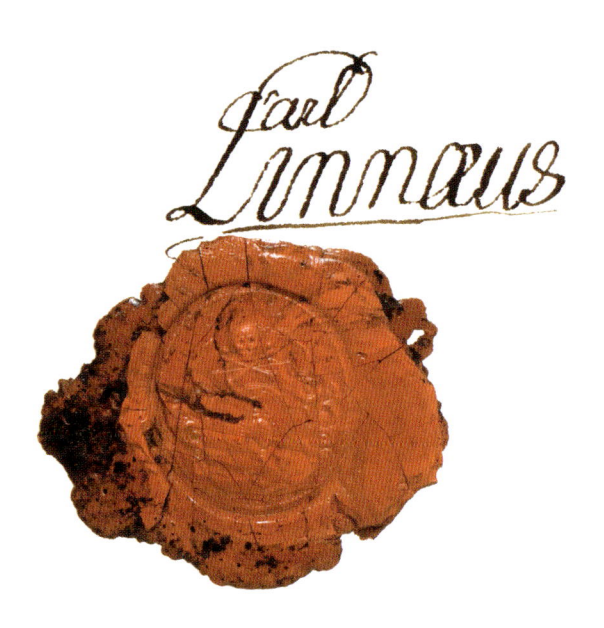

9 June 2008

Dear Professor Linnaeus,

This letter may surprise you from several quarters. We haven't met, but I felt compelled to write to you on the occasion of your recent birthday. Through your published works, influence and legacy, it feels like I am, nevertheless, communicating with an old colleague and friend. I hope you'll excuse this impertinent familiarity.

I am a newly inducted and proud Fellow of The Linnean Society of London. This august body was founded in 1788, 10 years after your death, by Sir James Edward Smith and others to commemorate your contributions to, and cultivate the science of, natural history. As a young scientist in an antipodean land, I admired from afar the Linnean Society and its rich history, publications and collections (including most of your own, acquired by Sir James and subsequently purchased by the Society, still securely curated and actively researched).

The year 1788 was also significant for me in that it was the date when the country of my birth was first colonised at Sydney Cove by the British due to the influential urging of your eminent younger colleague and natural historian Sir Joseph Banks. It is my great privilege to follow in the latter's footsteps, leading for a time the Royal Botanic Gardens, Kew, which remains a pre-eminent global force in botanical science and horticulture.

Of your many accomplishments, I often reflect upon aspects that have profoundly influenced the world since your time. Above all, perhaps your greatest contribution was your strong and consistent advocacy for the use of binomials as a convenient short hand to convey the names of organisms. Almost three centuries after you first published on this matter, we still apply generic and specific epithets.

Portrait of Sir Joseph Banks during his tenure as President of the Royal Society by Thomas Phillips RA (1809)

94   *Letters* to *Linnaeus*

A fanciful rendition of James Edward Smith's acquisition of Linnaeus' collections and their transport to England

This is a remarkable and practical legacy because, as you aptly pointed out: "each object ought to be clearly grasped and clearly named, for if one neglects this, the great amount of things will necessarily overwhelm us and, lacking a common language, all exchange of knowledge will be in vain."

It may amuse you to know that we still are describing new species of plants at the rate of around 2,000 a year. The process you set in train of listing and describing all the world's organisms is by no means complete. It has become one of the grand ventures of western science, involving thousands of practising taxonomists over three centuries. Binomials have, indeed, become the currency of biology, enabling a myriad of useful, enjoyable and intellectually stimulating intersections between people and other organisms. I would like to draw your attention to a recent editorial[1] that summarises our present state of affairs in this regard.

Although a relatively minor contribution compared with yours, I have had the pleasure of being involved in the ongoing discovery and collaborative description of some 300 plant species new to science, mostly endemic to the Southwest Australian Floristic Region. You will not be familiar with this biologically fascinating part of the island continent, other than reference to New Holland in the annals of Dutch maritime exploration. It has emerged as second only to the Greater Cape Floristic Region of South Africa among the world's temperate districts richest in plant species, with some 8,000 native species known, 50% of these endemic, and an estimated 14% still to be described scientifically.

Indeed, because other similar corners of the world still remain poorly explored for plants, a global list of species remains elusive, despite valiant attempts since your time to summarise what is known. However, 2010 has been set as a target for a modern world checklist in the Global Strategy for Plant Conservation, and there is recent evidence from collaborative interactions among leading botanic gardens that a first draft in electronic form of such a checklist may be possible to meet this deadline.

Of course, botanists will continue to debate and differ on the circumscription of genera and species, so the exact number of world species is never likely to be a point of universal agreement. However, we have recently achieved great advances in resolving part of this problem through the power of molecular phylogenetic analysis of DNA sequence data. I fear you will be somewhat troubled by the underlying philosophy embodied in this modern terminology.

It has become clear from the brilliant and scientifically consistent work of Charles Darwin, who lived a century after you, that species are not immutable, as you believed so fervently, aside from the influence of hybridisation. They change through time, evolving from common ancestors in a continuous magnificent tree (or multi-stemmed shrub) of life. We now have an understanding of how the process of natural selection works on inherited variation, transmitted through self-replicating DNA, causing differential reproductive survival among individuals within populations. Moreover, with modern molecular techniques, we can sequence mutational changes in the structure of DNA, and apply methodologies enabling reconstruction of probable phylogenies of any group of organisms. All species, ours included, are part of this superb lineage through time, dating back to the emergence of life a billion years after the 4.6 Ga origin of our planet. This great story is something that you, with your religious beliefs, may find difficult to contemplate. But such is the power of science as a body of independently testable knowledge that few today other than religious fundamentalists have not come to accept the close approximation to truth that Darwin's concept of evolution through natural selection embodies.

I have elaborated a little on this subject only to highlight that the power of taxonomic classifications is immeasurably improved in terms of predictability about relationships by adopting a genealogical view. Hence we are no longer solely reliant on similarity of morphological features, as you were, in deciding upon how to assemble species into genera, genera into families, etc. Shared ancestry has become the dominant taxonomic criterion, and we have learnt some surprising things about the mutability of organisms and the convergent evolution of similar structures from this perspective.

This seemingly academic issue in systematics, of course, has profound practical ramifications. It does and is causing changes in the names applied to organisms, much to the chagrin of those countless millions of people who use binomials or generic epithets routinely. But it also has introduced a degree of scientific rigour to the naming of things undreamt of in your time, when only the opinion of illustrious experts such as yourself carried the day (except, of course, when another illustrious expert preferred a dissenting and alternative scheme, presenting an agony of choice and a diplomatic quandary to users of taxonomy). Having a predictable classification based on the robustly inferred phylogeny of life has enabled great leaps forward in comparative biology, and in the exploration of attributes of living organisms useful to people, such as new foods, medicines and other pharmaceuticals, timbers, etc.

I note that one of your students Olof Rydbeck wrote a poem in 1762 expressing concern at the loss of forests and meadows occurring in your native Sweden, pitying those who would live 100 years hence in a world bereft of "an inheritance given to us". This challenge to your view of "God's endless larder", of nature existing in inexhaustible abundance for human use and domination, must have been somewhat troubling. Yet, as I have found, students have a way of asking the naïve question and seeing things in a prescient way which all old professors should heed and celebrate. Rydbeck's lament has now become a global chorus. Deforestation alone is contributing a fifth of global carbon dioxide emissions today, more than the world's prodigious transport systems, warming the planet and changing climate at an unprecedented rate. No longer is wild nature perceived as endless and indestructible. Rather, it has become threatened, with no corner of the world removed or safe from the impact of people. Without a fundamental shift in values and in the use of Earth's biological inheritance, we run the risk of losing the very diversity that so much inspired your life and career.

Biologists today could learn much from your skilled use of advocacy and diplomacy in advancing their issues. We are at a point in history where scientists more than ever before must engage with people from all walks of life to ensure we care for other organisms. For the first time, more people now live in cities than in rural areas globally, making the need to stay in touch with nature all the more urgent for a sustainable future. The diversity of plant life in particular is essential to the habitats that form our world, and is vital for our own well-being. The ongoing erosion of wild vegetation is degrading the quality of life for billions of people and prejudicing the drive to eradicate poverty. Plant life must be conserved, repaired and restored if climate change is to be moderated and humanity is to have a tolerable future. If we save threatened plants and restore habitats, we could improve the quality of life for people. The future will be brighter than present forecasts suggest. You would perhaps find this fundamental change in perspective on the natural world the most challenging of all if you could see where we are today.

In conclusion, Sir, while things have advanced considerably in taxonomy since your time, there are few scientists in history who have had the essential aims and tools of their life's work carried on by future generations over three centuries. My congratulations to you for this fine achievement. If you were alive today, I would have enjoyed meeting you, discussing the immensely variable and interesting diversity of life, debating how best to prosecute the science of their classification and seeking your advice on how to influence decision makers in whose hands rests the fate of other organisms and the very quality of human life.

Yours sincerely

Steve Hopper

Professor Stephen D. Hopper FLS
Director
Royal Botanic Gardens, Kew

Notes:

[1] Hopper, S.D. 2007. New life for systematics. *Science* 316: 1097.

Dear Professor Linnaeus,

The winds of change that have come about since you lived 200 or more years ago have seen some rapid advances in the subjects of systematics and taxonomy.

Do you remember those days of late summer when violent thunderstorms came about? Did it ever occur to you to consider what the composition of the lightning might have been? Well, it is stuff called electricity and its capture and the ways of manipulating it have been exploited to the max, and so the start of the 21st century has been called the age of information technology. The composition of lightning is infinitesimally small particles called electrons, which have the ability to flow along metal.

That discovery took a while coming, through the work of such famous scientists as Alessandro Volta, Humphry Davy and Michael Faraday. They eventually realized that electricity was a natural force which could be captured. They also discovered that various substances such as rubber and plastic—solid compounds that could be collected from plants or oil from inside the earth—would not conduct electrons. Therefore, the electrons could be captured in such a way so as to put wires made of copper or zinc through non-conducting rubber and plastic. The technology was very powerful and from about 1800 to 1880 electricity became better understood. Since the capture of electricity there have been many tools built that utilise the energy in electricity, heat, light, sound and vision. So, with the invention of electric lights, heat generators, the ability to record music, voices and sounds and even to make and project moving pictures through television has allowed the world to become a very rich place since your time. Theatres, musical halls and clubs can all work in the dark

Electric light fitting in the port on the Río Madre de Dios in Puerto Maldonado, Peru

with great effect and resolution through amplification of sound, light and vision.

But, that is not all. The 20th century has seen great strides in the development of radio, the travel of sound through space, from any spot on the globe to any other spot on the globe. Earth is like a ball in space that travels around the sun in one year and spins on its own axis every 24 hours. These days we can even hear sounds from other galaxies and see the galaxies through powerful telescope and radio systems. But, to my mind, the greatest achievement of the 20th century is the development of powerful electronic systems called computers. Today, everyone in the western world has a computer, or access to one, and things are greatly improving in less wealthy parts of the world, such as South America, India and Africa. By the end of this century I predict that everyone will have access to a computer, with the prospect of gaining infinite knowledge through an integrated network and the ability to communicate with anyone in the world—as long as you know their e-mail address.

A man of your imagination can see what this means for systematics and taxonomy. No longer will we need to keep our information on index cards or volumes of paper but be able to store the information electronically in things called databases. Databases keep information in very tidy filing systems so that the first main effect has been to allow all names to be recorded in one place, and for the development of dedicated websites on the network for all kinds of information, from medicinal plants to every known insect. Anyone should, in theory, be able to access any of the databases and get the precise information they require from the international source. But there is more– since your time there have been many efforts made to improve on the way we sought out groups of related organisms. You were the great pioneer behind the binominal nomenclature system that is still pride of place today, but the ways in which groups within groups are now sorted is by their natural relationships. There has been a succession of methods of how to determine what can be called a natural group from overall similarity to genealogy, the latter being currently in vogue. There was a brilliant entomologist called Willi Hennig who lived in Germany and who devised a whole method called Phylogenetic Systematics. Many others have contributed to his technique and we now have very powerful computerised techniques to sort out so-called natural groups. I would cite people such as Stephen Farris and David Swofford who have worked on cladistic techniques and have provided us with the software to undertake electronic analyses on our modern computers. My portable computer is so light I can take it with me anywhere. I am often seen undertaking natural group analysis whilst riding the train from my house to my office in the middle of London. The only reason

I go uptown to work these days is to see the actual specimens of the groups of plants upon which I work, and to browse our library for previous published analyses or for determining the correct names of things. You will be pleased to know that this still figures highly in these endeavours especially since the publication of a fabulous book by the Linnean Society of London and the Natural History Museum, London called *Order out of Chaos* by a very unassuming character called Charles Jarvis. If you were alive today you would have it as bedside reading, the book being such a wonderful homage to your craft.

So my distinguished fellow, I can say that you are alive and well in all known natural history museums and botanical gardens. Rest in peace Carl—you are still one of the greatest scientists we have ever had and your work holds a great place in the 21st century. I hope you enjoy reading this collection of essays as much as the contributors have enjoyed writing them.

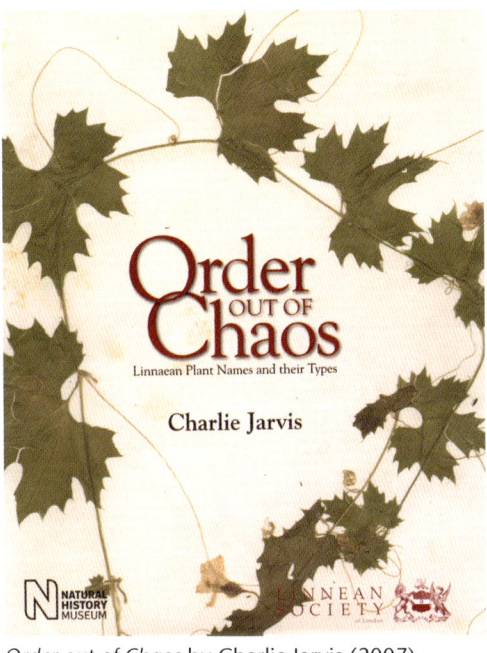

Order out of Chaos by Charlie Jarvis (2007) details the typification of all Linnaean plant names

Christopher John Humphries
Natural History Museum,
London, England.

Carl Linnæus

Pehr Kalm (1716–1779) was one of Linnaeus' students who travelled and collected in eastern North America from 1748–1751. His letters were long and convoluted, even Linnaeus sighed over their wealth of detail.

He writes that he [Kalm] has arrived in Norway and that even a coward at sea would have been happy with the crossing, so calm was the sea [Linnaeus was a famously bad sailor]. They have also been lucky as they have had no encounters with pirates, who are common in the area and take everything from their victims, even their clothes. . . . He complains about the cost of his correspondence . . .

Kalm says he has found lodgings with a family where the lady of the house speaks Swedish, but the 4 daughters do not; being women however, they will never keep quiet so he will have plenty of opportunities to improve his English and they have promised that after a fortnight with them he will be able to manage on his own without help. . . . He is reluctant to use Linnaeus' letters of introduction until his English is better. He feels he could always use Latin, but the English pronounce it so badly. He cannot understand them, nor they him, even though they speak the same language [Latin]. Kalm worries about Canada being the "most appalling and cold country under the sun".

May 31, 2008

Dear Professor Linnaeus:

I hope that it would not dismay you to learn, following the tercentenary of your birth, that you have become famous in some circles for something you undoubtedly never intended. That is, your very interesting taxonomy of *Homo sapiens* 'subspecies'—classified according to the four elements (air, earth, fire and water), the four corners of the Earth (north, south, east, and west), the four humors (black bile, blood, yellow bile, and phlegm), and the known continents (America, Europe, Asia, and Africa)—has become a textbook illustration of the nefarious effects of cultural presuppositions on scientific inquiry.

Before getting to that unintended consequence, though, let me say that you are to be congratulated for your other contributions to the 'science' of human classification. For one thing, you dared to challenge religious authorities by classifying human beings among the animals. For another, your perspicacity about human biology (which I will get to later) helped you avoid using the term *race* in your taxonomy and refrain from delineating the boundaries between groups. Finally, you resisted the temptation to rank *Homo sapiens* subspecies, even though your belief in a great chain of being could have justified creating such a hierarchy. Indeed, your penchant for detailed classification basically precluded such stratification, according to Winthrop Jordan, since it was not possible to produce a single hierarchical ranking of all living creation.

Some of your contemporaries, as well as several of your followers, did not resist that temptation, even as they proclaimed their allegiance to *monogenesis*, or God's creation of humankind as a single but varied species. For example, your protégé, the monogenesist Johann Blumenbach, argued in the 1790s that his own European lineage, which he dubbed *Caucasian* after the especially attractive women of the Caucasus Mountain region, was the original and most favored type of human. You did share something else with Blumenbach, however, and that is what led to the unintended consequence I mentioned previously. Whereas Blumenbach's distaste for darker skins and other non-European attributes and his association of beauty with the female sex were rather obvious expressions of his cultural presuppositions, yours were somewhat subtler, though equally present.

Let me explain. In the 10th edition of your *Systema Naturae*, published in 1758, you described the four major human subspecies as American (meaning Amerindian), European, Asiatic, and African. Each category represented a continent and compass direction. Each also represented one of the four humors—Americans were "choleric"; Europeans were "sanguine"; Asiatics were "melancholic"; Africans were "phlegmatic." Each group also had a distinct skin color and source of behavior and personality—Americans were red and "*regulated* by habit"; Europeans were white and "*governed* by custom"; Asiatics were yellow and "*governed* by opinions"; Africans were black and "*governed* by caprice." In addition, each group had a unique type of hair—for Americans, it was black and straight, for Europeans, it was thick and "abundantly flowing" (no color specified); for Asiatics it was simply black, and for Africans, it was black and "frizzled." Though the criteria for those qualities were consistent, your European perspectives and tastes are evident in the specific descriptions.

Europæus. β. albus, sanguineus, torosus.
Pilis flavescentibus prolixis. Oculis cæruleis.
Levis, acutissimus, inventor.
Tegitur Vestimentis arctis.
Regitur Ritibus.

Asiaticus. γ. luridus, melancholicus, rigidus.
Pilis nigricantibus Oculis fuscis.
Severus, fastuosus, avarus.
Tegitur Indumentis laxis.
Regitur Opinionibus.

Linnaeus' annotated entry for the European and Asian 'varieties' of humans in his own copy of *Systema Naturae*, edition 10 (1758)

Even more revealing of your differing preconceptions and standards of judgment for each group, however, were other less consistently framed observations. For example, you described Americans as "obstinate, erect, free," and painted with fine red lines, while you defined Europeans as "gentle, acute, inventive, and brawny." Did that mean Americans were uninventive or that they could not be measured on that criterion? And why not say something about European bodily adornment? If Asiatics were "severe, haughty, and covetous," and Africans were "crafty, indolent, negligent," and anointed with grease, did that mean Asiatics were therefore un-greasy and un-crafty? Or did it mean they could they not be judged in those categories? If they weren't indolent, were they therefore energetic? You don't say. And why did Africans' "tumid" lips and flat noses evoke comment, while you said nothing about other lips or noses? Why was African skin singled out as silky?

The sense of randomness increases in your *Monstrosus* category that followed the other four (I'm omitting mention of your *Homo sapiens ferus*, or wild children found in the forest). Apparently the *Monstrosus* category encompassed people with deformities as well as mythological creatures that you believed actually existed (in 1763 you would add troglodytes, satyrs, hydras, and phoenixes). The 1758 category included Alpine dwarves, Hottentots, Patagonian giants, the Chinese, and Canadians (probably Eskimos). It is unclear why the Chinese and Hottentots were extracted from the larger Asiatic and African categories respectively, but you seemed quite certain that the Chinese were noteworthy for their conical, long heads (were you counting the hats?) and the Hottentots for their "single-testicled, so less fertile" reproductive physiology. The dwarves made your list for being "active" but "timid" and the Patagonians for being large but "indolent." The Canadians were remarkable for their compressed and slanted heads. What might these observations reveal? The collection of traits reminds me of a list that Michel Foucault found in an ancient Chinese encyclopedia, which divides animals into: ". . . (a) belonging to the Emperor, (b) embalmed, (c) tame, (d) sucking pigs, (e) sirens, (f) fabulous, (g) stray dogs . . . (k) drawn with a very fine camel hair brush, . . . [and] . . . (m) having just broken the water pitcher . . ." among other incommensurable categories. According to Foucault, that taxonomy reveals "the exotic charm of another system of thought" as well as "the limitation of our own." To me, the encyclopedia list also suggests that all systems of classification can exude their own kind of "exotic charm." All are firmly rooted in the time and place of their creation, and even as they offer important data and theory, such systems reveal a great deal about their creators.

Thus, things that might not have struck you as remotely strange in your human taxonomy, even discounting the *Monstrosus* category, can seem quite 'exotic' to modern readers. For example, your deductive imposition on human beings of a predetermined concept of the natural order, in which continents, compass directions, and humors coincide, seems an odd choice over an inductive construction of categories based on close observation and data. Another oddity is your assumption that the primary subject in each of your subspecies is male. That is apparent from the mostly male-identified qualities you attributed to the four groups—"erect," "free," "brawny"—as well as from your single reference to women, under the category "Africans," which indicates that females were not part of your classification system unless specified. That singular female observation also suggests that you were primarily interested in women for their sexual and reproductive traits:

Natural order: 'The Anthropomorpha of Linnaeus', from *Amoenitates Academicae (IV: Anthropomorphae)* (1763), the dissertation of Linnaeus' student C.E. Hoppius

African "*Women's* bosom [is] a matter of modesty," you wrote; their "*breasts* give milk abundantly." We aren't told if they were crafty and indolent like their menfolk or whether they, too, were anointed with grease.

Male bias in your schema is also apparent in the adjectives you used to describe "Asiatics." In contrast to the other three groups, you not only found no positive characteristics to describe them but you also used terminology, such as "haughty" and "covetous," that was often used in 18th century parlance to describe inferior women. It is likely, therefore, that you, like many of your European compatriots, regarded Asians as "a feminine people," which was not meant as a compliment.

It might please you to learn that your four major subspecies of *Homo sapiens*, flawed as they may be, have endured longer than any other human taxonomy. They are still used by forensic anthropologists, politicians, and even physicians to denote the major human groups, supplemented by ethnicities in modern times. You might be less pleased to learn, perhaps, that others have turned that impressive (if somewhat misleading) legacy into a scientifically unsound system of racial classification. By the 19th century, various physical anthropologists and naturalists would identify from two to 64 human races. Even after the dust settled on smaller numbers, theorists had transformed the most scientifically robust aspect of your taxonomy—its association of physiological characteristics with geographical areas—into permanent biological categories. As you perhaps intuited, that notion of biological race contradicts Nature's distribution of particular characteristics in clusters around the globe, reflecting the effects of environment and geographical isolation rather than fixed racial identities. Indeed, it turns out that there are no genes for race *per se* and that there is more genetic variation *within* so-called racial groups than there is *between* them. An individual's ancestors can be most easily traced if they hailed from a relatively remote area in which there was minimal exchange of DNA with outside groups, which is not the same thing as tracing the DNA of race.

We are grateful to you for your courage and creativity as well as for the foibles of your human taxonomy, all of which teach an important lesson about objectivity and the role of culture in science. Today's scientists recognize that, like all human beings, they have been shaped by their cultures to promote their own interests, ask particular questions, and interpret the answers. In other words, scientists themselves are factors in scientific inquiry. Combating the risks to explanations of the natural world posed by that reality requires replicable experiments, communities of interrogators, and other strategies. Thank you for contributing to humankind's discovery of and solution to that problem. And a belated Happy Birthday!

Sincerely,

Sally L. Kitch
Director, Institute for Humanities Research
CLAS Humanities Professor, Women and Gender Studies
Arizona State University

Carl Linnæus

September 2008

Puerto Maldonado, Madre de Dios
Peru, South America

My dear professor,

I do wish you were here with me—you would scarcely believe it. I have travelled to Peru to attend a national botanical congress, imagine, a gathering of all Peruvian botanists in one place! There are scores of young men and women (you would approve of this I know) all discussing plants and how they are classified, how to identify them correctly, all the things you love. They even do as you did, and bring plants for one another to look at and help identify—you would be so pleased at how widespread and vibrant the science you so loved has become!

Peruvian students at the XII Congreso Nacional de Botánica, Puerto Maldonado, Department of Madre de Dios, Peru

You will recall you only knew of Peru via plants sent by the unfortunate de Jussieu brother (you remember, Joseph, the one who lost his mind and became a "martyr to botany" in South America, who sent seeds to Bernard who sent them to you!), but today, Peruvian botanists are hard at work documenting the wonderful flora of their own land. Peru, because of its enormous variety of habitats, is one of the Earth's megadiverse (such a word, I hear you say!) countries; the coast is dry desert, the mountains rise to above 5,000 metres in elevation and the eastern part of the country is the start of the great Amazon basin, whose area almost equals that of the United States of America, where Pehr Kalm collected such wonderful things. It is too bad that both your students who came to this continent never brought you the riches you expected (or did you expect this place to be so diverse?); poor Pehr Loefling died, and you didn't think much of Johan Dahlgren—but I hear that recently some of his diaries show he worked harder than people thought. In your day you thought there were only about 10,000 species of plants, today we estimate there are probably about 400,000 (we know of 350,000 already!), with more being discovered each year. South America is the richest continent in terms of plant species, and here botanists, like those attending the congress, discover new species every year. These New World tropical lands are a treasure house for plant richness; most of the plant species on Earth grow here. We all try to use your sensible principles to name our finds, but I am afraid the number of species is so many that we are using a huge variety of names you might not approve of fully—I am going to name a new *Solanum* in honour of a distinguished Peruvian botanist, that you would approve of I am sure.

I have seen your *Solanum peruvianum* and *Solanum montanum*, they both grow in a strange vegetation formation called 'lomas', meaning hills. Here on the coast of Peru, which is normally a dry desert, fogs collect in their winter, and with that little bit of moisture, a luxuriant vegetation can grow, but only for a short time. Every so often, a phenomenon known as 'El Niño' occurs—so called because it used to come around Christmas time—where more rain falls and the desert truly blooms. But today, because the climate is changing (mostly due to man's upsetting the economy of nature), and the rains come too often, or not enough. I heard of a species that has been wiped out by a huge wave triggered by an earthquake—did you ever think of species you knew disappearing from the face of the Earth?

People speak now of a biodiversity crisis—a crisis because it is feared that many plants and animals may disappear before we even give them names. This might not matter to many, but as you so rightly said in your *Philosophia Botanica*, "If you do not know the names of things, the knowledge of them is lost too." But this is more important than the 'mere' giving of names; we human beings depend utterly upon the rest of life on Earth, even though we often are unwilling to admit it. Who knows what effect there will be of all the plants in the lomas never occurring again, of the entire Amazon rainforest being cut down and planted in crops? We still know so little about how this wonderful planet works (you would, I fear, be surprised that there are more worlds than this one, but not surprised that not a single other planet has supported life). So what can we do? It is often tempting to throw up our hands and complain that we are too few, there is too much to do, or that it is just too difficult. But I think that we are weak if we do that—I see the young botanists here in Peru, with no libraries to speak of and few resources, cheerfully going about the business of describing and documenting the flora of Peru—what a sterling example!

We botanists (it really is the king of sciences) are fortunate in some recent developments at an international level (imagine many countries coming together and deciding to do something about the decline in diversity—mind-boggling). Almost 15 years ago the world's leaders came together in another South American city and agreed to understand and protect biodiversity—all of life on Earth. The convention they

Solanum peruvianum L. (Solanaceae) from Linnaeus' herbarium (LINN 248.17)

Huge trees being taken from the forests of the Amazon basin

signed is known as the Convention on Biological Diversity. As with any such agreement there are problems, red-tape and inactivity, but as part of the CBD we botanists have a plan for action to save the earth's plant species called the Global Strategy for Plant Conservation (GSPC—you see, we love our acronyms!). How we do this depends upon us, working together, across borders. The first task of this plan is to create a working list of all known plant species—a kind of *Species Plantarum* for our own century. I do hope we do as well with our task as you did with yours. Another task is to understand the threats to the survival of all plant species—a monumental task. But you know, my dear professor, I do think we can do it. It is a task that will never be done, as you well know, but what an accomplishment it will be when we are able to say we have established a baseline against which to monitor change in the centuries going forward.

I do wish you were here to see how big and exciting botany has become, and to relish the enthusiasm with which these young students are approaching the study of the enormous diversity of plants in Peru. The congress is being held in the town of Puerto Maldonado, deep in the Amazon basin, in a beautifully green and lush part of the country, with enormous diversity of plant species. I must end now, and apologise for sending few samples for your perusal. My own love of plants is always kindled when I reach these wonderful tropical forests, so I am sorry, but I must take a few hours away from hearing about botany to go out in the field and discover plants for myself!

Yours respectfully and in haste,

Sandra Knapp

Carl Linnaeus

Richard P. Korf

Emeritus Professor of Mycology
Department of Plant Pathology & Plant-Microbe Biology
Plant Science Building, Cornell University, Ithaca, New York 14853–4203, USA
Tel.: 607–273–0508, Cell: 607–280–5645
Fax: 607–273–4357, Email: rpk1@cornell.edu

1 August 2008

Prof. Carl Linnaeus
In Heaven (or, God forbid, in Hell)

Dear Carl,

I hope you will not be offended by my use of your first name, but in America it is permissible to address one's elder by first name if the relationship is close. In our case, we have never met, but your writings and your spirit have been with me from my first encounter with them as an undergraduate student studying mycology. Your invention of the system of binomial nomenclature made you my hero.

You have doubtless been both amazed and confused by the developments of rules governing the naming of plants and animals, and the development of Codes of Nomenclature, all specifically based upon your brilliant system of binomial nomenclature. These days, perhaps even more than say 50 years ago, attacks are being made on the systems that have worked so well since your time, with a bewildering new array of suggestions for naming based not on observation but on numbers and sequences of patterns deduced by chemical analysis!

A mushroom drawn by Linnaeus in his Lapland journal

From your vantage point, I cannot help but wonder how you see these new ideas. Yours were radical ideas in the 1700s. They have lasted two and a half centuries, still revered by the great majority of botanists and zoologists. In my lifetime I have seen more than a half dozen 'new' systems come (and go!), and I have no doubt still more to come, all to disappear as inefficient means of communication, all pale by comparison with the hierarchical system that is based on your work. You remain my hero, whose breadth of knowledge and brilliant analyses have been the foundation of all biological research. Without you at our side, we would have floundered about in a sea of confusion.

Here's hoping you keep well. Drink at least eight glasses of water a day, and bask in the sunshine.

Forever in your debt,

Richard P. Korf

Richard P. Korf
Mycologist

Getting back on track

Dear Carolus,

May I apologize beforehand that I write you in such a profane language, but English is the language in which scientists are supposed to write these days. Not that English is in any way particularly apt for scientific description and argument, rather less so. Despite being one of the richest languages in respect of sheer number of words, it is surprisingly deficient in descriptive terms. Describing a few new species of scarabs right now, I am searching for words that do not exist and wish I could write in French or my mother tongue, German. Well, English became the language of science merely by accident[1], because it was the language of the most successful winning party of the last global war (we have had a couple of them so far). But this is not why I write this letter today.

The reason why I bother you is that, finally, I can tell you great news. In zoology we might soon be as advanced as you were exactly a quarter of a millennium ago. When you wrote your *Systema Naturae*, at least in the later editions, you compiled all names of known organisms in one work. Everybody could access the complete catalogue of the known animal and plant kingdom and even the minerals in one place. Since then, we have made enormous progress in zoology and botany having described and named probably one and a half million more species. As a result, we lost track.

Although all zoologists, botanists and microbiologists have followed your binominal system of naming species, we no longer know how many species are described and which names are already used. Admittedly we have catalogues and lists available, but all of them are incomplete and full of errors. You may ask: Is there no central authority recording and regulating all these names? Yes, there is for the regulating part. Dealing with an ever increasing flood of new names, over a century ago the zoological community decided that some general rules were necessary to deal with naming and names. Actually it was you who first introduced rules for naming genera and species over 150 years earlier; I mean your *Fundamenta Botanica*, later expanded as *Critica Botanica*[2]. Again, you were ahead. With thousands of your successors naming species anywhere in the world according to your binominal system but following their own little idiosyncrasies, the need for a general, globally acceptable set of rules became inevitable. The International Commission on Zoological Nomenclature was founded in 1895[3] and is still in existence. It consists of 20–30 commissioners from all over the world. Your compatriot, Doctor Sven Kullander, the ichthyologist from the Museum in Stockholm, is a member. The Commission elaborated the *International Code of Zoological Nomenclature*[4] which has currently reached its fourth edition. It is a huge, complicated semi-legal book that does not attract readership. Don't worry; we are currently working on a more easily accessible fifth edition that might pave the way to what you had achieved in your time: a register of all animal names.

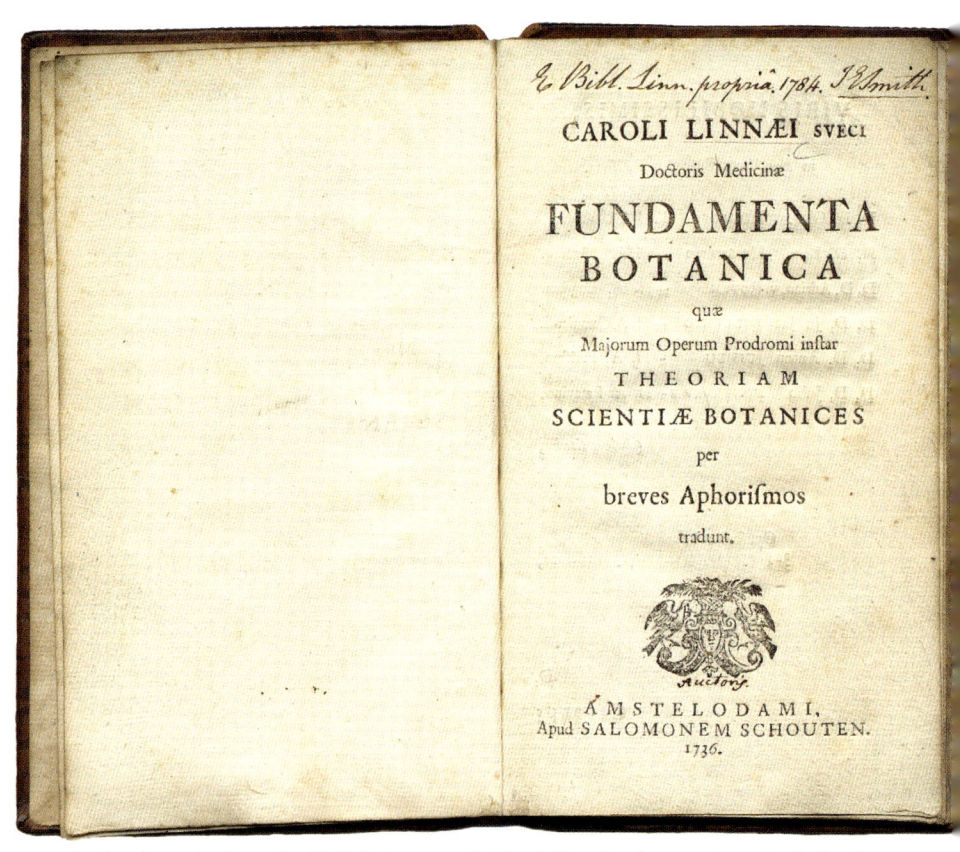

In *Fundamenta Botanica* (1736) Linnaeus laid out his rules for naming plants for the first time; later he expanded them in *Critica Botanica* (1737)

Although we have rules for how to name animals and plants and in zoology even a Commission responsible for those rules, we do not know all these names. If I talk to normal people (I mean non-scientists) about how many animal species have been described and how many more are waiting to be discovered, very often I am confronted with astonishment and disbelief that there is no central register for animal species. What would a library of over a million volumes be if it had no central catalogue of its holdings? This is our current situation. The botanists even voted against a central register for plant names at the International Botanical Congress in St. Louis[5] (there are scientific congresses in America now). No author has voted against library catalogues though.

The International Commission on Zoological Nomenclature plans to propose registration for all zoological names[6] as a voluntary exercise. The technical framework of the registry is currently developed and will be available to the scientific community in 2009. It looks promising so far. Editors of several journals have expressed interest in registering any new names their journals publish. Authors might find a central registry useful and register their new names if their editors do not. You introduced binominal nomenclature. It was not mandatory in your time, but since it was an ingenious and useful concept it gained acceptance rapidly. Useful concepts often gain acceptance if not forced upon people. I am confident that the 'Republica Zoologica' will join forces and work on an authoritative register of all animal names. The new 'Systema Naturae' will be magnitudes larger than yours was. My apologies, I do not intend to disdain your achievements; nothing was further from my mind. You wrote more books than even the most prolific living scientist. You developed the common language of binominal nomenclature. Not many people can claim having developed a communication system that is still in common use after 250 years. You did. We can only work on making your tested and proven communication system more efficient, and this is what we are trying to do. I shall inform you about further developments. I trust that you approve of our efforts. If you have doubts, please do not hesitate to write me in clear words. Coming from you, I expect these words to be in Latin, and they will be very welcome.

Your devoted servant

Fr. Scabiosus

Franciscus Scabiosus
alias Frank-Thorsten Krell, FRES, doct. rer. nat. Univ. Tubing.
Cur. Ent. Mus. Nat. Sci.^{que} Denv.
Soc. Curiae Gen. Nom. Zool.
Soc. plur. soc. sci. Americ., Germ., etc. etc. etc.

The dung rolling beetle *Scarabeus sacer* L. (Scarabeidae) was the sacred Scarab of the Egyptians

Notes:

[1] Kaplan, R.B. 2001. English—the accidental language of science. Pp. 3–26 in: Ammon, U. (ed.) *The Dominance of English as a Language of Science. Effects on Other Languages and Language Communities*. Berlin: Mouton de Gruyter.

[2] Linnaeus, C. 1737. *Fundamenta Botanica*. Amsterdam; later Linnaeus, C. 1737. *Critica Botanica*. Leiden.

[3] Melville, R.V. 1995. *Towards Stability in the Names of Animals. A History of the International Commission on Zoological Nomenclature 1895–1995*. London: International Trust for Zoological Nomenclature.

[4] International Commission on Zoological Nomenclature 1999. *International Code of Zoological Nomenclature. (4th ed.)*. London: International Trust for Zoological Nomenclature.

[5] Greuter, W. 2004. Recent developments in international biological nomenclature. *Turkish Journal of Botany* 28: 17–26.6

[6] Polaszek, A., Agosti, D., Alonso-Zarazaga, M., Beccaloni, G., Place Bjørn, P. de, Bouchet, P., Brothers, D.J., Earl of Cranbrook, Evenhuis, N., Godfray, H.C.J., Johnson, N.F., Krell, F.-T., Lipscomb, D., Lyal, C.H.C., Mace, G., Mawatari, S., Miller, S.E., Minelli, A., Morris, S., Ng, P.K.L., Patterson, D.J., Pyle, R.L., Robinson, N., Rogo, L., Taverne, J., Thompson, F.C., Tol, J. van, Wheeler, Q.D. & Wilson, E.O. 2005. A universal register for animal names. *Nature* 437: 477.

Dear Dr Linnaeus,

I am writing to congratulate you on the 250th birthday of your magnum opus *Systema Naturae* (10th edition). It was an amazing feat in itself to develop a system for the naming of all life known in the middle of the 18th century. What is even more extraordinary is that a system based on simple physical similarity without an underpinning theory of the biological *processes* that generate the diversity of life on Earth has managed to stand the test of time.

So how has your system fared in the intervening years?

Countless students around the world have learned about your binomial system, from humble schoolrooms to exalted university lecture theatres. Generations of schoolchildren have attempted to master the basic idea of the binomial system and have taken this with them unknowingly into adult life. Children, barely old enough to read a decent piece of writing, will enunciate "*Tyrannosaurus rex*" or "*Diplodocus carnegiei*"; they don't yet know that they are using your naming system but they will. Of course, your starting with naming plants was a really good idea, the expanding horticultural world in the 19th and 20th centuries with its legion of devoted amateur enthusiasts, especially in my own country, has

Fossilised skull of *Tyrannosaurus rex* Osborn (Tyrannosauridae), a binomial name familiar to many from childhood

produced an ardent band of followers. It is even possible to go into huge open air shops, which we now call 'garden centres' where one is able to purchase plants from all parts of the world for exceedingly less than a week's labour, to hear all manner of people using your very system as the *lingua franca*.

Your system is used in all our popular media, from newspapers, with which I think you will be familiar but perhaps not with the scale of all classes of people that read them, to our new means of mass communication. Millions of people listening or watching this new communication medium will hear binomens used. It is just incredible how ubiquitous your system has become and one of the very few parts of science that is used by all sectors of society. Our politicians pass laws which embody names based on your system. There are not many scientists that have such a broad legacy. In fact, I can't think of many of your contemporaries that are as well known by the general populace as you are, and certainly none in what we now call the life sciences.

I'm sorry to report however that your disciples have made the business of nomenclature rather more complicated than you may have intended. They are certainly as learned as you were in the vagaries of ancient European languages, which is somewhat of an obstacle to the true internationalisation of nomenclature in the modern scientific world where millions of people use languages of other origins. But, perhaps, what is most disappointing is that we now have different rules for different groups of organisms. We do stick to the basic idea of a binomen but how we generate and handle names is different, not for scientific reasons based on biological reality, but because of history and an unfortunate fractionation of the natural philosophers. As an optimist I hope we will unite these different systems of rules to reflect the growing realisation of the unity of life and the drawing together of natural philosophers in what we now call interdisciplinary scientific research. I am not sure how many see the irony in this return to an 18th century holistic approach to natural history within the guise of modern exploratory science. We need to undertake this reunification quickly to ensure the subject remains vital and is able to exploit the huge opportunities presented to us to explore the living world and communicate using new technologies for handling information.

Your system is very pragmatic and simple, which I believe is one of the main reasons it has survived so extraordinarily well, and why it is in such widespread use, as I hope I've indicated already. However, it is increasingly difficult to reconcile your classification system with the remarkable discoveries we are making on how the diversity of life is actually generated. How life really works. We now know that life's diversity was not made as a static set of forms, your species, in which the observed variation in nature was an imperfect expression of some essential quality. No, it's so much more exciting than that. Populations of organisms are able to change in response to each other and their environment, and by an extraordinary range of mechanisms. This process of change we call evolution. The idea was advanced by a Mr Charles Darwin who published a revolutionary paper 100 years after your 10th edition, which very much changed how we see the natural world. He is probably one of the most widely known biologists of all time.

The process of evolution means that at any one point in time some lineages will be so distinct from one another that they can be fairly easily circumscribed and fit reasonably well into your naming system. Given that we know many species in a rather superficial manner, as we only have a few observations and physical appearance to go on, we can only hazard a guess (an educated and structured guess I would like to suggest) at variation and kinship. The binomen is a crude expression of the relationships between closely related (not necessarily similar) organisms but, as I've said previously, it's pragmatic and a fair first approximation. The real problem comes when lineages are not so easily circumscribed because we either have so much more material to study or because we have very powerful tools by which we can examine Nature.

The Sacred Lotus *Nelumbo nucifera* Gaertn. (Nelumbonaceae) resembles a Water Lily, but is not related

Put simply, as lineages are diverging, they are neither one 'species' nor another. Your binomial system means we cannot easily record this true biological state of affairs. We are forced into an unnecessary straight jacket. This is not comfortable nor does it aid communication of the variability and complexity of life as we are increasingly coming to know it. Populations are radiating, fusing, permanently or transiently. We need to communicate this. As a mathematician fluent in the language of set theory might put it, your system is based on complete sets, naming an entity as *Aus beus* means that it cannot be in any sense part of *Aus ceus*; there is no ambiguity. We need to operationalise the concept of fuzzy sets where populations can have a definable membership to more than one group. It is a reflection of their change, their development over time. This ambiguity reflects our increasing knowledge of the process of speciation for multicellular organisms, but as we come to know more about the microbial world, with their profligate shuffling of genes (another clever discovery) and indefinable boundaries of 'species', your system will become increasingly more difficult to apply.

Discovering and describing the variety of life on our planet still remains a major challenge, so long after your great step forwards. We, the international scientific community, have managed to delimit and describe some 1.7 million species but this is a far cry from the number of species we expect to occur in nature. At the rate of about 10,000 new species described each year we will take until 2050 to describe only twice the 1.7 million. If there are as many species out there as we expect (perhaps 3 million + or even more) then we will not complete the task in the next 250 years. This is a great disappointment, not only to meet the scientific challenge of describing all life but also because we currently face a 'biodiversity crisis' in which we are seeing a mass extinction of species to rival those that have gone before in geological history.

Time is not on our side if we wish to deploy our expertise in naming organisms to assist in slowing the rate of biodiversity loss. We need a radically different approach to our science. The current approach is inadequate to meet needs, so simply ramping up productivity using existing nomenclatural and publication tools will not suffice. There needs to be a fundamental change from the basic assumption that we need to describe everything in order to understand the scope and the processes that generate biological diversity to only using formal description in those taxa where a formal name is essential. The emerging technologies for handling information that I have alluded to, collectively called biodiversity informatics, can associate different kinds of information with unique identifiers that do not require a formal name for a taxon. So we shall not lose the ability to communicate about Nature, but will probably increase our ability to be faster and more accurate.

Fortunately for us, new technology has enabled us to look at the previously unknown biodiversity in the micro-world of invisible and barely-visible organisms. It is here that perhaps the greatest growth in our subject will be made in the future. 'Molecular' tools to look at nucleic acids such as DNA, one of the greatest scientific discoveries of the last century, are enabling us to make discoveries possible at a remarkable scale and speed that we can only marvel at. No longer are we confined to describe the world of microbes that we can grow in the laboratory or see under our new electronic microscopes (it's a microscope that does not use light! Strange but true). Using these molecular tools, we are now beginning to take an ecosystem approach to describing the diversity of micro-organisms based on extracting nucleic acids from a particular habitat and trying to deduce the number of unseen organisms that are there (its called metagenomics). This is not anywhere as precise as the standard methods of delimiting taxa, and there are quite a few theoretical and practical problems to overcome, but it is giving us an insight into true biological diversity that we could only speculate on before.

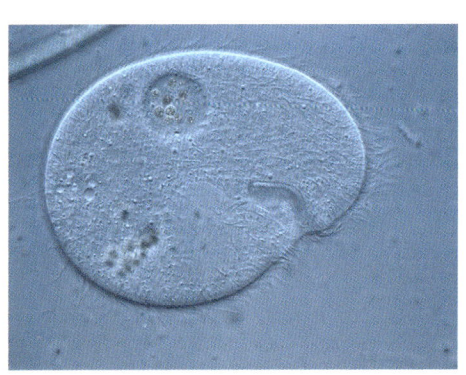

Unicellular eukaryotes like the ciliate *Plagiopyla* were not known to Linnaeus but today's techniques allow us to explore their diversity

To address the challenge of describing life on Earth we are beginning to change from an 'artisanal' approach, with which you will be quite familiar and frankly, it is surprising that it has continued for quite so long, to an 'industrial' scale approach taking advantage of new technology. Industrialisation is another major shift in society from its humble origins in the 18th century which brings together human and mechanical resources to transform production. It's not just about lots of people being in the same place doing the same thing but, together, working quite differently. Using this new approach to organise ourselves, we are beginning to use massive DNA sequencing and other forms of data capture, large scale output and analysis of data and high throughput identifications of specimens to have a broader view of biodiversity. This will increasingly require us to work in large international teams rather than individually. It will be demanding but exciting.

The business of naming organisms, working out their relationships to one another and the environment, has never been in such a promising position with a plethora of theoretical and practical tools and means of communication. The challenges are great but the outlook looks promising so long as we work in different ways.

So, my dear Dr Linnaeus, what of the future of your system? Will we be using it in another 100 years? I believe we will be using your basic binomial system because it is pragmatic and simple and easily understood by fellow scientists and the 'public', but I don't think we will be using your system alone. We will augment it with more subtle naming methods in which we are able to reflect the true complexity of life based on the mechanisms by which biological processes generate diversity (rather than ways of handling data). Thus we will likely have different naming systems based not on taxonomic group and established social networks, as at present, but on the true biology or state of knowledge of organisms.

Your system for the naming of life has been remarkable, enduring. I am sure you will be as proud of its legacy as we are using it.

Richard Lane,
Director of Science,
Natural History Museum,
London, UK.

Carl Linnæus

Honourable Sir

Printed in Linn. Corresp.
v. 2. 559.

I Hope your goodness will excuse the Liberty I have taken in addressing myself to you, as it proceeds from a knowledge of your superior Merit, and your exalted character in Natural History. I have been employed some years past, by the King of great Brittain in collecting of Plants for the Royal Gardens at Kew, my researches have been chiefly at the Cape of good Hope, where I had the fortune to meet with the ingenious Docter Thunberg; with whom I made two successfull journies into the interior parts of the country; My labours have been crowned with success, having added upwards of 400 new species plants to his Majesties collection of living plants, and I believe many new Genera.

I expect soon to go out on another expedition, to another part of the glob, to collect plants, for his Majesty, and if I should be so fortunat to discover any thing New in any branch of Natural history I should be happy in having the honour of communicating it to you. I had the pleasure of seeing Mr Sparrmann at the Cape, and received from him a parcel of seed, which he collected in the Southeren Islands, which I now send you, I could not presume to send you any cape Plants as I presume Dr Thunberg has sent you every kind that he hath collected which are much the same with mine

I also intended sending you a collection of the S.t Hellena Plants but that I shall refuse untill your Ports are open for Shipping.

The inclosed specimen I think is a new Genus, which my worthy friend M.r Thunberg had a great desire of giving the name of _Massonia_, and honouring me with a mark of his friendship. But notwithstanding the good will of D.r Thunberg, and many other Botanical friends I have declined receiving that honour from any other Authority than the great Linnæus, whom I look upon as the Father of Botany and Natural history, in hopes that you will give your Sanction. I am sorry that the leaves is not perfecter but it was the only Specimen that I had. I shall only take the liberty to give a Discription of the Root & Leaves, the Flowers being perfect I shall refere to your better judgement

Radix bulbosa, tunicata, subglobosa, diametro sesquiunciali.
Folia duo, radicalia, ovata, subrotundo, acuminata, glabra, lævia, carnosa, nervosa, nervis immersis, palmaria; adspersis maculis purpurascentibus.
Scapus brevissimus, sinubus foliorum quasi immersus, teres glaber. Hab in campis elevatis in Rogafeld distr.

I have seen another Species with narrow leaves which flowerd in Kew Garden which I shall send you next opportunity I shall add no more but am with great esteem Your
Most Obedient Humble Servant, Fran.s Masson

London December 26 1775

Direct to M.r Francis Masson at M.r James Lee's Nursery Gardener at Hammersmith near London

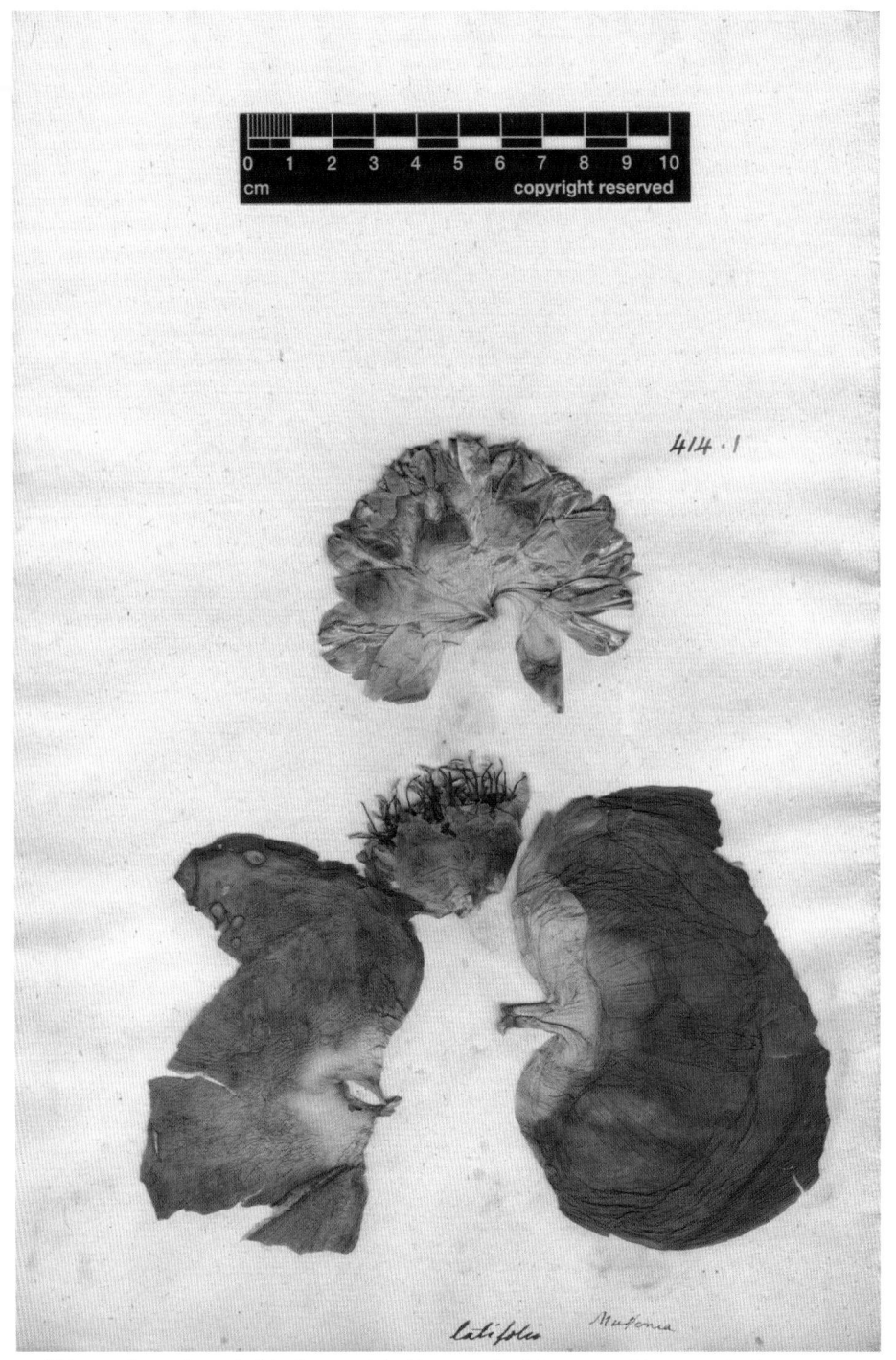

This specimen is probably that sent to Linnaeus by Masson and is annotated "*Massonia*" in a shaky hand by Linnaeus himself. It was used to describe *Massonia latifolia* by Linnaeus' son.

Francis Masson (1741–1805) was a collector in South Africa, and later in the Caribbean. He travelled with Linnaeus' pupil Carl Thunberg in South Africa, and was one of the great early collectors for the Royal Botanical Gardens at Kew. In this letter he writes to Linnaeus from London on the 26th of December 1775, only a few years before Linnaeus died. Masson was staying with Mr James Lee at his nursery in Hammersmith. The plant in question was not described by Linnaeus but by his son; Thunberg did indeed name the genus *Massonia* in Masson's honour.

"Honourable Sir,

I hope your goodness will excuse the liberty I have taken in addressing myself to you, as it proceeds from a knowledge of your superior merit and your exalted character in Natural History. I have been employed some years past by the King of great Brittain in collecting of Plants for the Gardens at Kew, my researches have been chiefly at the Cape of Good Hope, where I had the fortune to meet with the ingenious Doctor Thunberg, which whom I made two successful journies into the interior parts of the country. My labours have been crowned with success having added upwards of 400 new species ~~plants~~ to his Majesties collection of living plants, and I believe many new genera.

I expect soon to go out on another expedition, to another part of the globe, to collect plants for his Majesty and if should be fortunate to discover any thing new in any branch of natural History I should be happy in having the honour of communicating it to you. I had the pleasure of see Mr. Sparrmann at the Cape and received from him a parcel of seeds which he collected in the Southern Islands, which I now send you. I could not presume to send you any cape Plants as I presume Dr Thunberg has sent you every kind that he hath collected which are much the same with mine. I also intended sending you a collection of the St Hellena Plants but that I shall refere untill your Ports are open for Shipping.

The inclosed specimen I think is a new Genus, which my worthy friend Dr Thunberg had a great desire of giving the name of Massonia *and honouring me with a mark of his friendship. But notwithstanding the good will of Dr Thunberg, and many other Botanical friends I have declined receiving that honour from any other Authority than the great Linnaeus, whom I look upon as the Father of Botany and Natural history, in hopes that you will give your sanction. I am sorry that the leaves is not perfecter but it was the only Specimen I had. I shall only take the liberty to give a Description of the Root & Leaves, the Flowers being perfect I shall refere to your better judgement.*

[Description in Latin]

I have seen another species with narrow leaves which flowered in Kew Garden which I shall send you next opportunity I shall add no more but am with great esteem

Your

Most obedient Humble

Servant,

Francis Masson

Direct [reply] to Mr Francis Masson at Mr James Lee's Nursery Gardener at Hammersmith in London."

Dear Quentin and Sandy,

I understand that through some fortuitous combination of good connections at the top and access to high-end time- and space-bending physics gear, you have a way to communicate with Linnaeus. What a coup! Only you could have pulled this off! As you know, I'm new on the job of ExecSec at the ICZN, but I would love to tell the big man what he set in motion, and my first impressions of the modern 'Linnaean apostles'.

Cheers,

Ellinor

London, 2008

Most respected Professor,

Of the approximately 100 billion people who have ever lived, very few have made a lasting contribution of any kind. Of that rarified fraction who have really altered the human world around them, an even smaller proportion have changed the way all of humanity sees and communicates about the rest of the living world. Arguably, this is the most profound influence any person could have. And this, Professor, is what you have done.

I'm writing to give you an update on the incredible process you started, and to give you a sense of who the modern day Linnaeans are.

Two hundred and fifty years ago you commandingly suggested we quit with the nonsense of either fully descriptive names for plants and animals, or names that are meaningful only in a local context, and that we use simple, unchanging names that mean the same thing to people the world over. Two names are enough for starters, the first is like a person's family name, providing immediate context among other similar kinds, the second modifying the first to provide uniqueness, like a given name. The genius was in the simplicity, and obviousness, of this idea. And if, by some fluke, scientists gave different names to the same thing, which would eventually lead to communication breakdown, we follow a rule: dispense with the Johnny-come-lately name and use the oldest name. A pragmatic, if somewhat autocratic, solution to what clearly was a growing problem in an age of discovery yet poor communication among scientists.

These elegant rules have served us well, but new problems have cropped up in the maintenance of an unambiguous set of names for the living world. More rules have been added, now recorded in 'Codes' for each major taxonomic, or disciplinary, branch of the living world. It turns out that it is hard work to follow the rules. For animals, a fat book with golden letters on the cover presents the *International Code of Zoological Nomenclature* in 18 chapters of 90 articles, a code of ethics and a list of recommendations. In the name of clarity it is written with tedious, repetitive language. Zoologists often dread the study of this text, as they tend to be people of an independent, even iconoclastic, mindset, yet they attempt to follow the rules essentially voluntarily. The modern practice of nomenclature is (largely) a remarkable example of international cooperation for a shared goal of clear communication in biology. The Code should keep us out of nomenclatural trouble. Even so, mistakes get made, unpleasant subterfuges are enacted, fierce arguments break out, old and forgotten names get resurrected and sunk. The landscape of zoological nomenclature is surprisingly dynamic.

Times have changed since your *Systema Naturae* presented 4,819 species and higher taxon names; there are now 15,000–24,000 new nomenclatural acts published *each year*. Clearly, a judicial body is needed for name usages that run afoul of the 'law'. I have recently taken on a job at the fulcrum of this task, as the enabling arm of the ICZN. The 'C' in ICZN stands simultaneously for the International 'Code' and 'Commission' of/on Zoological Nomenclature, working towards '*sense and stability in zoological nomenclature*'. The job of the Commission is to keep the language of zoology fit for purpose in the modern age—that is the '*sense*' part of the motto—yet keep it from shifting with facile winds of scientific or social fashion—the '*stability*' part of the creed. The Commission attends to individual cases, but focuses on those that require an exception to the rules presented in the Code. It is not an enforcer; it does not seek out the hooligans of nomenclatural practice, it only deals with corrections presented by working taxonomists, and even then, it only attends to the problems that can't be corrected by a good taxonomic publication. However, the rules of the Code deal with the robust (as in documented, verified, and tied to a standard) presentation of *names*, not taxonomic decisions. Specialists on an animal group remain free to sit in judgment on the inclusiveness or narrowness of species, genus or family group depending on the data they see as most informative and on their perceived reality of taxonomic gaps. Since your foundational work, *Systema Naturae*, the concept of a type specimen has helped to nail down a physical standard for a name, and thus emancipate the relationship between names and the groups to which they refer. Here's a quick sketch to help you see this:

Nomenclature · Taxonomy

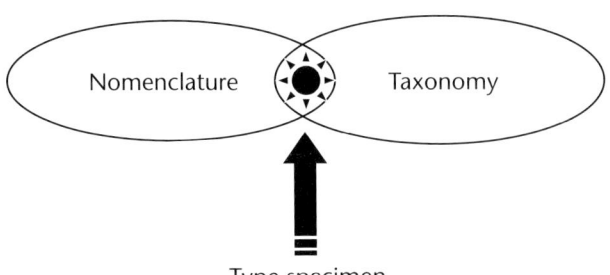

Type specimen

More importantly, the Commission makes major decisions on the future rules for zoological nomenclature—it is entrusted with updating the Code. Only nine years have passed since the last edition, yet modern practice has changed and problems with the previous document have been exposed through use, thus a revision is necessary. The most pressing question today is what qualifies as verified, archived published evidence of nomenclatural work. This highly legalistic question is necessary as an increasing amount of our science is coming out on the modern medium of electronic publications (words on light screens, transmitted with formulae, not letters formed by ink on pages). The actions needed are so pressing that an amendment to the Code is being thrashed out now, with the aim of passing it as soon as possible. Recently, well-meaning taxonomists have created crises by publishing new nomenclatural acts in sources that have no paper copy or physical archive. Under the current Code, their nomenclatural acts are not available. The Commission has drafted an amendment that addresses this problem, but you might be surprised that the world now aims to be very democratic, so this is open for public comment for a year before it can be passed.

You might be curious—who are these people who voluntarily put themselves at the forefront of a job that requires an enormous input of time, careful consideration of legalistic details, all for decisions on names of often-obscure taxa? The decisions they take usually deal with topics outside the hurly-burly of daily life; they don't increase crop yields, make anyone rich or create works of artistic genius. Nomenclatural work doesn't even discover new taxa *per se*. When I started this job I was a little concerned that I was relegating myself to the company of doctrinaire obfuscators. But nomenclatural decisions are crucial to biological work; they spin the threads of the fabric of our science and have long-lasting consequences. The practitioners are people with a selfless commitment to establishing robust foundations for zoology. It is a rare privilege to work with scientists of such dedication!

Who are the approximately 25 distinguished zoologists who serve on the Commission? Well, dear Professor, my first experience with them indicates that they are perhaps more like you than if you'd selected a group of acolytes yourself. They are much more trailblazers than foot soldiers. I have just had my first chance to meet many of them, and regular correspondence has filled in the gaps with the rest over the course of the past half-year. Commissioners come in all the various permutations you might have imagined yourself, had you come from different circumstances; from contrasting cultures, with assorted taxa to reflect your scientific curiosity. Like family, they are you, but sometimes more so, in ways you didn't expect. Some are adventure hounds, some are bibliophiles, some are theatrical, some are quietly effective, some are vaultingly ambitious, some are deeply focused on minutiae, some are bombastic, some are self-mocking, but most are a combination of these things at different times. Some are men, and even some are women! Perhaps you presaged all of this in that splendid portrait of yourself wearing a native Sami woman's costume from your adventures in the field!

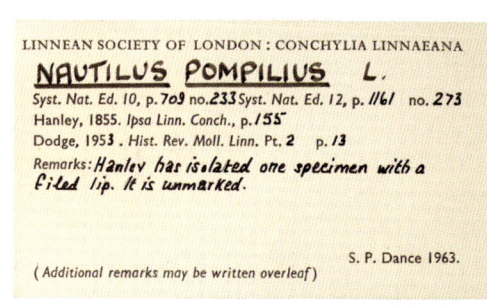

Linnaeus' specimen of the Chambered (Pearly) Nautilus *Nautilus pompilius* L. (Nautilidae) from the collections of the Linnean Society of London, with a label written by shell expert S. Peter Dance in the mid-20th century

How would you find the company of the Commissioners? Undoubtedly you would empathize with their basic passion for nomenclature, their love of biological diversity combined with a desire to bring order to the exuberance of the natural world. You would recognize their unusual intellectual combinations of almost savant-like skills in recognition of taxa and yet an ability to step back for a larger picture in setting the agenda for a whole discipline. You might be perplexed by some of the discussions that consume their attention, like electronic publication, which is a wholly modern problem. You would certainly pitch in with vigour in other current arguments that are right up your alley such as the fervent arguments about whether rules of ancient Greek and Latin should be enforced in current names. I imagine you would suggest a solution in short and magisterial order, but would be disappointed that the Commissioners would not simply adopt your proposal, either *de facto* or *de jure*. They will not let pass an opportunity to dissect a process down to its essential components, test them against each other and moreover, as they are representatives and not sovereigns of what is now a very wide world of practicing taxonomists, they will seek input from the global community of zoologists. Where once there was one of you, there are now many, and even they are representatives of yet more. These would be your friends and worthy opponents, doing the work you started, bringing nomenclature into the modern age.

Finally, you would be astonished and delighted with ZooBank, our internet solution to the logistical problem of a registry and catalogue for 1.8 million animal names. The principle of an instantaneously updated, always current, *Systema Naturae* available to scientists in all corners of the globe would need little explanation and would certainly receive your hearty approval. It will be, essentially, Linnaeus in every home and office, ready at a keystroke! You have provided us with a legacy as relevant today as it was when you published it, in its definitive 10th edition, 250 years ago.

With appreciation and respect,

Ellinor Michel
Executive Secretary
International Commission on Zoological Nomenclature

Carl Linnaeus

CALIFORNIA
ACADEMY OF
SCIENCES

55 Concourse Drive
Golden Gate Park
San Francisco, California 94118
www.calacademy.org

Dear Carl Linnaeus,

It is a strange pleasure to write to you. Strange in light of your death 230 years ago, so there can be no reply. And strange because so few write letters anymore. We do communicate continuously with those in distant places, but cursorily with hasty e-mails and phone calls. The pleasure of writing lies in the chance to address what I see as misperceptions that contemporary readers may have about your scientific efforts. Writing also provides a chance to reflect on some changes in the discovery and ordering of biological diversity, and to note a few things that have changed relatively little.

What I know about you is filtered through the translations of your writings in Latin or Swedish to English and the various interpretations of your views by the historians and scientists that have written about you. Certainly there has been mythologizing and simplification of your views which changed considerably during your life. However, there can be no mistaking your passion for facts, your keen eye for detail and pattern in nature, and your prodigious energy, which drove you to seek the natural order among entities from the species of roses to human morals and diseases.

Because the binomial, hierarchical system you implemented was so broadly adopted, giving us many of the names and some of the biological language we still use, the scientific and cultural inheritance you bestowed on us is simultaneously valuable and inadequate. I doubt this would surprise you. As our understanding of the origins and the particulars of life's diversity has improved, the system has required revision. With the thoroughly unfair benefit of hindsight, I will start by noting a few widely perceived shortcomings.

First, you did not see clearly enough the unifying theme of evolution, meaning common ancestry for all life forms, though you might have. At least one naturalist of your era did, Benoît de Maillet. Though you did integrate the idea of common descent into some of your species descriptions in *Species Plantarum*, these were inconspicuous and buried (though clearly explained) within your large body of work. You failed to extend these evolutionary observations and engage the social constrictions placed on the scientists of your day. Perhaps as the son and brother of pastors and as an academic at a time when church leaders controlled university appointments you chose to sublimate your unconventional views for practical reasons. Your soaring, biblical rhetoric, for example, asserting "All species are certain diversities of form which the Infinite Being created so in the beginning; which forms according to immutable laws of generation produce always their like . . .", showed clearly your profound religious commitments which likely predisposed you to miss this singular opportunity.

Second, the use of the sexual parts for taxonomy and classification in plants provided a good bit of success; however, extrapolating this sexual system to non-plants was much less successful. As it turns out life is too variable, and frequently too asexual, for such an approach. And third, despite your efforts to develop consistent methods for the taxonomy of species, genera and families that could be broadly applied, your approach, a mix of objectivity and subjectivity, was really not up to the task. You may take comfort in knowing, however, that we are still debating the best approach to taxonomy; and distinguishing subjectivity from objectivity remains difficult.

Now, to elaborate a few points in your defense. As noted above you did come to see that some species likely arose naturally, in more recent, post-Creation times—a view supported by your observations of species distributions in the wild as well as hybridization among plants in your gardens. As you say, in several cases there can be no doubt that one, "has before one [a] new species, produced by cross-fertilization . . . This, therefore, lays a new foundation, on which natural scientists should be able to erect a great building. From this it would seem to follow that the many species belonging to the same family were in the beginning a single species, and that they have since arisen through such cross-breeding." But it was not just hybrids that arose as distinct new species. You noted for example, that *Thalictrum lucidum*, a meadow-rue, ". . . is not so very distinct from *T. flavus*. It seems to me to be the product of its environment." And in discussing *Achillea alpina*, a yarrow, you said, ". . . may not the Siberian mountain soil and climate have molded this out of *A. ptarmica*?" Your implication of environmental differences as influencing species origins, heretical in your own time, is quite reasonable by modern standards.

Importantly, you clearly saw the fatal problem this presented for your plant classifications, noting, that because ". . . it is possible for new species to arise within the plant world; . . . families with dissimilar fructification organs can have the same origin and character; [thus] . . . the natural classification of plants [is] exploded." For these reasons you abandoned in the latter editions of *Systema Naturae* the theory that no new species arise, and the hope for a simple taxonomic method.

No doubt this was painful, at least for a time, but this willingness to admit prior errors and then work to resolve them is consistent with the best aspects of science, and today's evolutionary biologists, including taxonomists, see you as a kindred spirit. Also recognizably independent in your thinking, you were willing to push back, at least in your private correspondence, against the pronouncements of religious authorities impinging on your expertise. For example, after the Lutheran Archbishop of Uppsala accused you of 'impiety' you wrote, "It is not pleasing that I place Man among the Primates, but man is intimately familiar with himself. Let's not quibble over words. It will be the same to me whatever name we use. But I request from you and from the whole world the generic difference between Man and Ape, and this from the principles of Natural History. I certainly know of none. If only someone might tell me just one! If I called Man an Ape or vice versa I would bring together all the theologians against me. Perhaps I ought to, in accordance with the discipline of Natural History."

Further, in advocating scientific study of the natural world and in railing against superstitions you wrote that without scientific study, ". . . Demons of the forest would hide in every bush. Specters haunt every dark corner. Imps, gnomes, river spirits, and the others in Lucifer's gang would live among us like gray cats, and superstition, witchcraft, black magic, swarm around us like mosquitoes. . . . The sciences are thus the light that will lead the people who wander in darkness." Despite the

Achillea alpina L. (Asteraceae), the Chinese or Siberian Yarrow (LINN 1017.13)

Achillea ptarmica L. (Asteraceae), or Sneezewort (LINN 1017.11)

apparent superstitions you maintained on some topics, none can doubt the priority you gave to science where evidence was available. I only wonder if you would be surprised at the degree of public misunderstanding and the frequent political abuse of science in modern times. Science is our only productive method for understanding nature, yet as a species we are still easily misled by non-scientific approaches. As an open system disregarding cherished beliefs, science can and does upset ideologues with other ways of knowing.

Here are a few biological updates that I think you would find interesting. It was your passion to find the natural order inherent in life's diversity, and this work is far from complete, remaining one of the most consequential and exciting intellectual challenges that we face. Indeed, our ignorance expands as we first encounter entire groups of previously unknown organisms. Your work, in keeping with 18th-century Natural Theology, was science in the context of religious contemplation, a way to understand the methods and mind of God. Though there are many scientists today who maintain a faith in God (leaving the need for evidence aside), the social pressure is now much reduced for professing faith in God, and allowing for attribution of natural as opposed to supernatural causes for life's origins and change. And yet the relevance of understanding the history and mechanisms generating life's diversity (evolution) could hardly be greater in importance for understanding the human condition. Evolutionary biology provides empirical understanding of our origins, behaviors (individually and in groups), cultural development, and our relationships to other life forms, both historically and ecologically, including our reliance on other species for human survival.

Discovering common descent as the organizing principle for life's diversity has helped us to recognize many more biological entities than in your day; and not just distinctive populations or 'varieties' within species of plants and animals known to you. Ongoing discovery of microscopic organisms, bacteria and viruses, are radically expanding the numbers of distinctive taxa that we would like to understand and name, but whose diversity of forms we have yet to circumscribe. Some of these microbes cause severe or fatal disease as a byproduct of infecting people, and this understanding of their evolution and the providing of appropriate names is a valuable component of disease prevention.

Comparative research on genomes (the material of inheritance) of these microscopic, asexual life forms has led us to recognize many instances of transfer of genetic components among distant relatives, generally known as lateral gene transfer or reticulation. Thus common ancestry is atomized, to some degree, with swapping of organismal components, which in turn yields a network of relationships among chimeric organisms. Some suppose that this invalidates discovery of any natural hierarchy for biological diversity. In my view, this is too pessimistic. The hierarchy will be recoverable in many instances but not others, and reticulation only highlights the challenges involved in discovering the pattern of historical change. Discovering patterns of lateral gene transfer can only be done in a phylogenetic context, and arguably this endeavor is contiguous in purpose with your earliest efforts in seeking the natural hierarchy among life forms.

So, it is true that we have extended your interest in the hierarchy of species to an interest in hierarchy among components of the genomes, particularly, but not only, for microbial organisms, and I can tell you this is not a look into any well-ordered supernatural mind. Rather it is a look into a changeable sea of genetic elements whose movements and actions often appear random and opportunistic. We have the well supported impression of genomes as pools of potential information that are optimal in some ways but decidedly suboptimal in others, with much excess material, messy fragmentation of functional units and even warring components.

Having absorbed the notion of chimeric, asexual organisms, I imagine you are worried about 'species' as recognizable entities in nature, and you would find that concern shared by many today. Even for sexually reproducing organisms, the continuous processes of evolutionary change, including the ongoing origins and extinctions of particular genome components and particular organismal lineages carrying those components make recognition of species a challenge. In my own view, species remains a useful designation, and phylogenetic methods for comparison among individuals, and for seeking uniquely shared traits, offer the most powerful approach to species delineation. Indeed, phylogenetic hierarchy is increasingly seen as the information we would like to communicate in our classifications. Ah, but of course there are politics involved.

Politics, in terms of differing views on which traits are key in recognizing taxa and categories of taxa, arise because there is no standard regarding degree of difference in morphology, physiology, life history, behavior or niche breadth, that can be broadly and objectively applied to guide taxonomic category delineation, particularly for higher (highly inclusive) groups. For example, flight has been deemed sufficient to elevate birds to the category of class, but not so for pterosaurs or bats.

However, I think the real breakthrough for taxonomy and classification, other than phylogeny itself, is our improving ability to estimate time, in millions of years, for the branching points on a phylogeny.

Time is a physical constant reliably quantified by radiometric methods (unknown in your day) that does provide an objective metric. Time passes at a regular and independent rate in comparison to rates of change in organismal features, and there are no subjective disagreements about the relative importance of this or that millennium as there are for traits enabling this or that form of locomotion, metabolism or replication. Further, passage of time can be measured and recognized, throughout life's history and across the tree for all life forms, whereas the histories of particular organismal traits, that might inform the ranking of categories, are less easily discovered, due to phenomena we know as convergence, selection and chance events.

Once the benefits of this are seen more clearly, and lingering antipathy for dating methods fades, I think this will be accepted as the single objective criterion for classifications. Having a timeline for lineage origins will improve our analyses of the processes giving rise to biological diversity and it will give us a better understanding of the history of life on Earth and may even help us in predicting the future of life, in ways yet unknown.

In discussing the lack of monetary reward for plant taxonomy you said, "Alas that so much toil should gain so humble a reward; and yet how welcome to the recipient such a reward should be! For even if knowledge of the true and original Tree of Life, which could have postponed the arrival of old age, is lost, the plants . . . shall always exhale the sweet memory of your names and make them more lasting than marble . . . For riches vanish, the most stately mansions fall into decay, . . . but the whole of nature must be obliterated before the genera of plants disappear and he be forgotten who held the torch aloft [in botany]." And it is still true that taxonomy as a career pays poorly, if at all, and its value is grossly underestimated.

You used the term 'Tree of Life' above in its biblical context—referring to the tree God planted in the Garden of Eden whose fruit bestows immortality. Current scientific use of the 'tree of life' phrase is to describe the single complex genealogy, simplified as a branching diagram, showing the pattern of common descent among all life forms. Despite these different meanings, I would like to think you might find some pleasing symmetry in them. The mythological Tree of Life in Genesis and the historical genealogy for all life forms linking humans with the earliest known life forms 3.8 billion years ago (and the potential future of life on Earth) does indeed provide an element of immortality as promised by the tree of mythology. There is no lack of poetry or awe-inspiring facts in our scientific world view.

Despite the lack of success for many of your projects (such as seeking to grow non-native plants like coffee, tea and rice locally in Sweden) those experiences were instructive nonetheless. You wrote, and more than a few current scientists will relate, "We poor humans, we toil and strive; we make slaves of ourselves, we begrudge our bodies rest at night; and this to acquire merit, to gain grace and favour, to make ourselves skillful, and all we gain in the end is hatred, disfavour, and grief . . . You, Gentlemen, who will outlive me, do not put it down to laziness if you hear nothing more of me, but put it down to the times and destiny . . . now you see in another what you may sometime have to taste yourselves." Buck up, Carl! Johann Wolfgang Goethe wrote of you in 1817, "After Shakespeare and Spinoza it is Linnaeus who has had the greatest influence on me." Many have forgotten the impact that you had, in bringing a systematic organization to the naming of life's diversity, based on keen observation of their natural features, and often with poetic style. But we recognize you still as a visionary and tireless pioneer in an ongoing and vital effort.

Oryza sativa L. (Poaceae): rice cultivated in Uppsala, Sweden (LINN 460.1)

To paraphrase a closing sentence of yours to fellow naturalist John Ellis, "Adieu, and may you live long in the Phoenix of our science."

David P. Mindell

Carl Linnaeus

My dear Linnaeus,

Your name and your work have accompanied me along many important segments of my activity in zoology.

Throughout 40 years of study and research, I have often wondered about the reasons for your lasting success and popularity. I soon found an answer in the work of an old countryman of mine, Abbot Giuseppe Olivi, who in 1792, just 15 years after you left this world, published an excellent account of the marine life in the Adriatic Sea, where he declared that he whole-heartedly adopted your classification and your nomenclature because your system of living beings was the most sensible and complete among those available at the date and the language you used was the most suitable for communication purposes. With his *Zoologia Adriatica*, Olivi was eventually to contribute in a major way to establishing the Linnaean tradition in my country and I was proud to start moving my steps along such a noble path.

In those years I was passionately busy with insects, beetles and dragonflies especially, under the guidance of Milo Burlini, a knowledgeable amateur entomologist who at that time was completing a monograph of the Palaearctic species of the chrysomelid genus *Pachybrachis*. Milo had also a very good knowledge of our native flora. As a true Linnaean, when hunting for plants, he used to bear, hanging from his neck, a copper vessel with a lateral window through which he placed the specimens as he collected. That vessel was a good specimen of the *vasculum* you had described, dear Linnaeus, in your instructions to voyagers and collectors. A nice tool indeed, but by the time I became a student at the University of Padova (the *studium* from where Pietro Arduino corresponded with you between 1761 and 1765), it had been already replaced by much lighter, user-friendly plastic bags.

The mountains of Italy are rich in both plant and animal diversity (Alpi Apuane, Lucca)

Through Milo and through the books I used to identify my specimens I become soon acquainted with your name and your importance in animal and plant systematics, but I wanted to know your work directly, rather than through the indirect or second-hand citations. Happily, in the Municipal Library of my home town Treviso, northern Italy, I had the opportunity to move from the Italian prose of Giuseppe Olivi to the Latin pages of your *Systema Naturae*. Not those of the 10th edition, however, but those of the less rare and much bigger 13th, posthumous edition of your masterwork. Your Latin was quite dissimilar from the language of the classics I studied at the Lyceum, but it was not too difficult to grasp, both in the telegraphic diagnoses of individual genera and species and in the more articulated commentary sentences, but especially in the grandiose introduction to the whole work.

But let me go back to the reasons of your success. Why did people worldwide celebrate in 2007 the tercentenary of your birth much more solemnly than the tercentenary of Georges Louis Leclercq, Comte de Buffon? This was perhaps unexpected, if we consider the immense popularity enjoyed in the past by Buffon's *Histoire naturelle générale et particulière*, admittedly a work immensely more readable than your books, and not only because he always wrote in French rather than in Latin or in a less popular modern language like Swedish, as you did. Perhaps, your lasting popularity is mainly dependent on the fact that you effectively advertised, if not entirely invented, a couple of tools people needed in your times, and still need today, to organize, store and disseminate information about the diversity of living beings.

The first of these tools is, of course, binominal nomenclature. The other, possibly better appreciated by my colleagues in botany than by my closer fellow zoologists, is the standardization of terms used to describe the structure of plants and animals and, to a lesser extent, their biology. The fact that this second part of your encyclopaedic program has been less appreciated and furthered than the first part does not detract value from your efforts in providing a terminology that still offers a wonderful scope for research (and a never-ending challenge) to people compiling matrices of morphological data to be used in phylogenetic analyses. Forgive me, my dear *magister*, for using words that were not yet in use in your times. Words, somebody would say, which embody concepts too far away from your world view, and even contrary to it. But I do not share their concern. I think that this kind of comparison between the research programs of scientists from different centuries has little to be commended, not unlike the all too common fashion treating extant and extinct 'species' as strictly comparable things. But these are not problems of yours, or of your times.

Well, everybody acknowledges that the success of your works, *Species Plantarum* and *Systema Naturae*, is the main cause for the nearly universal use of the system of naming animals and plants we still call Linnaean nomenclature. This could be an adequate reason for keeping alive the memory of your scientific activity although, in comparable contexts, only science historians remember the name of the first author who systematically used the current set of symbols for the chemical elements, such as Au for gold, or Cu for the copper of which your *vasculum* was made. Your name might have elapsed from common awareness had the set of animal and plant names in use remained basically limited, since your times, to the approximately ten thousand entries in your standard works. Two and a half centuries of taxonomic history have demonstrated, however, that the number of species to be discovered and eventually described and named is enormously larger. At the same time, to accommodate this amazing amount of information, and names, in your *Systema Naturae*, one could, in principle at least, simply continue to scribble new names and descriptions on the page margins of a copy of your own works.

Vasculum used for plant collecting by Charles Robert Darwin

To be sure, this is not to say that the current views about biological classification are the same as yours. Nevertheless, many basic targets and tools of the taxonomist are not that different from yours, to the point that—were it not for the language used in diagnosing new (animal) species and a few minor technicalities—the basic works produced by taxonomists today would not be that different from yours. In other terms, the successful working tool you put in the hand of taxonomists in the second half of the 18th century set in motion a largely additive process, which continues uninterruptedly in the work of today's taxonomists. In a sense, what makes all active taxonomists your fellows is the personal contribution each of us can offer to this daily update of the *Systema*. And I dare to say that it is this continuity, much more than the simple use of the binomial nomenclature (a practice we share with a huge number of 'external users'), that actually keeps alive our interest in your works, despite the fact that the percentage of taxonomists knowledgeable in Latin is, for many reasons, increasingly vanishing.

I am upset by the thought of what might have happened around 1748, when Julien Offray de la Mettrie published his malicious *Homme-plante* where he ridiculed your sexual system of plants. I know how distressed you became, to the point that you decided not to publish a single line anymore. Fortunately, however, this was not your last word on the subject and later on you resumed working at your *Species Plantarum* monograph, thus eventually establishing on firm ground the botanical use of the binominal nomenclature and subsequently extended it, with the 10th edition of *Systema Naturae*, to the animal kingdom too.

Continuity, thus, of your work. But, to be sure, continuity *with modifications*, as that other great master, Charles Darwin, rightly highlighted as the unavoidable fate of all lineages. A continuity, from your beginnings down to today's taxonomy that has passed through both kinds of events we nowadays recognize in the history of plant and animal lineages through time, i.e., cladogenesis and anagenesis. Cladogenesis, in our studies, has been sometimes visible and abrupt, with the sudden formation of a new 'school' around a published manifesto, sometimes imperceptible and spread over a long time span, as happened with the slow but relentless increase of the differences between botany and zoology in their respective versions of Linnaean nomenclature. The tempo of anagenetic changes has been also unequal, spurred from time to time by major paradigm shifts in biology or by technical advances that improved our ability to collect, observe, describe, compare and eventually classify an impressively increasing sample of biological diversity.

Thus, irrespective of your obviously enormous popularity among today's taxonomists, is it actually fair to still call ourselves Linnaeans? I do not know the answer the majority of my fellow botanists and zoologists would give to this question. On the one hand, during the '90s of the past century there was an effort to provide a new bridge between the zoological and the botanical tradition in nomenclature, but the idea of a single *BioCode* to be applied uniformly to all kinds of living beings failed to come to fruition. On the other hand, dissatisfaction with the dependence of the 'Linnaean' taxon names on arbitrarily assigned ranks such as genus or family has spurred an effort to develop the 'non-Linnaean' alternative called the *PhyloCode*, whose chances of success are, at present, quite limited. Shall we really look for a universal language in taxonomy, or do we better accept some degree of pluralism?

All language is, at the same time, a tool for communicating and also a straightjacket for our thoughts. Scientific knowledge is increasing all the time and a frozen language such as the Latin formulae we still use, my dear Linnaeus, following your example, are obviously inadequate to express all the concepts we need to circumscribe in words. But this is not to say that 'your' names have become worthless. Arguably, Linnaean nomenclature is not suitable to name things of which you could not even imagine the existence, such as the hybridogenetic kleptons among the European green frogs, but the unprecedented number of new animal species annually added, in these times, to your initial world inventory in the 10th edition of *Systema Naturae*, and regularly provided with a 'Linnaean' name, is the most convincing witness to the unceasing value of your work. Let's feel free to introduce additional kinds of names, however, whenever we need to communicate about things other than those you wanted to serve by introducing the trivial names. Anyway, let's identify the needs first, and then look for a way to communicate about a new, specified set of things. Is this not the way you worked too?

While we continue citing (if not reading) your works, let's enquire about the reasons of your lasting success. This will be, to our benefit, your ultimate lesson.

Thank you very much, my dear Linnaeus.

Alessandro Minelli

University of Padova, Italy

Carl Linnaeus

Dear Professor Linnaeus:

Summer greetings from New Jersey! As I look out in my yard—I am writing this at home—your influence is everywhere, as many of the trees, like white and willow oak *(Quercus alba, Q. phellos)*, red cedar *(Juniperus virginiana)*, and flowering dogwood *(Cornus florida)*; shrubs, like mountain and sheep laurel *(Kalmia latifolia, K. angustifolia)* and shining sumac *(Rhus copallina)*; wildflowers, like teaberry *(Gaultheria procumbens)*, partridge-berry *(Mitchella repens)*, and bellwort *(Uvularia sessilifolia)*; insects, like Monarch butterfly *(Danaus plexippus)*; birds, like blue jay *(Cyanocitta cristata)* and American robin *(Turdus migratorius)*; and mammals, like eastern chipmunk *(Tamias striatus)* were first named and described by you over 250 years ago. All of these things in my yard having names coined by you are even more remarkable when one considers that you never visited North America.

Your influence extends well beyond my yard. Approximately one half of the plants that occur in the New York Metropolitan area were first described by you. I regret to inform you that many of these species, including Canada anemone *(Anemone canadensis)*, Canada milk-vetch *(Astragalus canadensis)*, butterfly pea *(Clitoria mariana)*, dewdrop *(Dalibarda repens)*, water lobelia *(Lobelia dortmanna)*, American milletgrass *(Milium effusum)*, basil bee balm *(Monarda clinopodia)*, prairie phlox *(Phlox pilosa)*, dogberry *(Ribes cynosbati)*, pod grass *(Scheuchzeria palustris)*, and buffalo clover *(Trifolium reflexum)*, can no longer be found from our area due to habitat loss and alteration. All we have left are your names and some herbarium specimens, such as those collected by your student, Pehr Kalm. Sadly, one of your favorites, *Linnaea borealis,* is also only historically known from the New York metropolitan area. Some plant species you named are rare throughout their range, including the pine barrens gentian *(Gentiana autumnalis)* and swamp pink *(Helonias bullata)*.

Helonias bullata L. (Melianthaceae), the endangered Swamp Pink of the eastern United States

On the upside many species you named remain quite common, such as Canada and seaside goldenrods *(Solidago canadensis, S. sempervirens)* and long-stalked and New York asters *(Aster dumosus, A. novi-belgii)*. However, some of our most common species in the New York area are originally from the Old World and are now displacing our native species. Most of the weedy grasses in my yard that were described by you, including hairgrasses *(Aira carophyllea, A. praecox)*, sweet vernal grass *(Anthoxanthum odoratum)*, junegrass *(Bromus tectorum)*, Bermuda grass *(Cynodon dactylon,* originally described by you as *Panicum dactylon)*, and bluegrasses *(Poa annua, P. bulbosa)* are native to Europe.

Your artificial sexual system of classification of plants is no longer in use. It was abandoned, not because of its language, objectionable to some, but because more natural systems, based on many more characters and currently accepted modern theories of biology, have been developed. These natural systems should not surprise you since you predicted they would be developed.

While your sexual system of classification has been abandoned and some of the species you named are in danger of becoming extinct, your rank-based system of naming remains safe. Indeed, the current codes of nomenclature for botany and zoology use your works, *Species Plantarum* (1753) and *Systema Naturae* (10th edition 1758–59), as starting points for their respective systems. Names published prior to these publications have no standing in modern biological nomenclature. Every generation there are attempts to replace your simple system of naming. Some have wanted to replace your names with numbers and others have wanted to abandon your formal ranks. None of these alternative systems has ever gained widespread acceptance as they fail to recognize people's preference to communicate in names and desire for the order that the ranks provide.

Your most lasting nomenclatural innovation was your *nomina trivialia* for species, or what we now call binomial nomenclature. This remarkably simple and wonderfully useful method of attaching an adjectival epithet to a generic noun has revolutionized the way we communicate about species. Polynomials (your *nomina specifica*) proved too long and cumbersome and uninomials inadequate to accommodate the millions of species that we now know to occur on Earth. (Your estimates of species numbers were seriously off, the tropics proving a lot more diverse than Lapland.) Binomials are just right and the current codes of nomenclature for animals, plants and bacteria—you were correct in rejecting the abiotic origin of these tiny organisms ("mushrooms for brains" you called its supporters)—use them for species names. In botany, we have a detailed system of abbreviations of authors of plants names, and you will be pleased to know that only you, L., were given a one letter abbreviation.

Some of your innovations have gone beyond the field of science and entered into popular culture. Everyone now refers to the medieval planetary symbols for Venus (♀) and Mars (♂) as the female and male symbols, respectively, thanks to your first use of them in this manner. And of course, 100° Celsius is hot and 0° is cold and not vice versa thanks to you and others who inverted the original Celsius scale, thus making 100° and 0° the boiling and freezing points for water.

Linnaeus used variants of standard male and female symbols in his Lapland journals

Getting back to taxonomy—the Lamarckian term we now use for formal classification—while you gave us an excellent system for naming the Earth's biodiversity and a wonderful head start with all of your descriptions of species, the task of naming all species on Earth is far from complete. Besides the task being much larger than originally thought, the field of biology has branched out so much that most biologists are no longer actively involved in the naming and describing of species. Some biologists today spend all their time studying just one of the species you originally described, such as mouse ear cress *(Arabidopsis thaliana,* originally described by you as *Arabis thaliana)* and corn *(Zea mays).* Sadly, many species are lost before they are ever named, and other times they are described only after they have been introduced from regions far away from their native ranges, such as a centipede *(Nannarrup hoffmani),* recently described from Central Park in New York. Nonetheless, we can all be pleased that the naming and describing of the Earth's biodiversity does continue.

Zea mays L. (Poaceae), known as Corn or Maize (LINN 1096.1)

Let us hope that someday all organisms on Earth are like those in my yard—named!

Sincerely yours,

Gerry Moore

Gerry Moore
Brooklyn Botanic Garden

Dear Carl Linnaeus,

I write to thank you for setting ichthyology on a secure footing by adopting the views of your friend Petrus Artedi, and in 1738 publishing his manuscripts after his untimely death at age 30 in 1735. His views, and yours, subsequent history has repeatedly proven sound, even now nearly 300 years after.

I note that in the first edition of your *Systema Naturae*, which you published at age 28 in 1735, you list—five genera of whales aside—36 genera in your Classis IV, Pisces: *Raja, Squalus, Acipenser, Petromyzon, Lophius, Cyclopterus, Ostracion, Balistes, Gasterosteus, Zeus, Cottus, Trigla, Trachinus, Perca, Sparus, Labrus, Mugil, Scomber, Xiphias, Gobius, Gymnotus, Muraena, Blennius, Gadus, Pleuronectes, Ammodytes, Coryphaena, Echeneis, Esox, Salmo, Osmerus, Coregonus, Clupea, Cyprinus, Cobitis, Syngnathus*. As you clearly state in your introductory *Observationes in Regnum Animale*, all of these derive from Artedi's *Ichthyologia*—his summary of the history of ichthyology, including his own many and careful observations and analyses.

I note that in the 10th edition of your *Systema*, published in 1758, you list the first five of these genera in your Classis III, Amphibia Nantes, adding a sixth genus, *Chimaera*. I regret that Artedi did not survive to advise you to maintain them in your Classis IV, Pisces, itself including 51 genera. There, your 22 additions are *Trichiurus, Anarhicus, Stromateus, Callionymus, Uranoscopus, Ophidion, Scorpaena, Chaetodon, Sciaena, Mullus, Silurus, Loricaria, Fistularia, Argentina, Atherina, Exocoetus, Polynemus, Mormyrus, Tetraodon, Diodon, Centriscus, Pegasus*. As you indicate in your citations, about one-half of these, too, derive directly from Artedi, and about one-half of the others from his synonymies.

I hope that you will be gratified to learn that two omissions in the 10th edition (*Osmerus* and *Coregonus*) were subsequently decreed, and are now universally admitted, to have been present there after all, in what we have chosen as the beginning of ichthyological nomenclature.

All of these 59 genera are universally recognized today, but we call them families. The reason is that we know so many more species of them. Your genus *Clupea*, for example, began in 1735 with four European species, but the equivalent modern group of clupeoids has more than 300 species, found in marine and freshwater, in all continents and oceans. Even so, all of your generic names persist but with a more restricted meaning evident mainly to the specialist. Your genus *Clupea* today includes only one, the herring (*Harengus*), of the four species you listed in 1735; the sprat (*Spratti*), shad (*Alosa*), and anchovy (*Encrasicholus*) are in genera of their own, and in the case of the anchovy also

158 *Letters* to *Linnaeus*

The Sprat was placed by Linnaeus in the genus *Clupea*, along with herrings, but today is known as *Sprattus sprattus* (L.) (Clupeidae)

in a family with some 150 allied species throughout the world. There are now many more than 59 families of fishes, about 400 in total, but in general they, too, are recognizable in the same way as your genera—at a glance, even in silhouette, by the experienced student. Such is not generally true for fish groups of any other rank above or below the family.

For this reason families are the basis for the practical teaching of ichthyology. And for this purpose, families function nowadays much as did the genera in your time—which is not surprising because they are, after all, the same entities in human perception. I know that you felt that a serious botanist should aspire to know all of the genera of plants, then amounting to several hundred. In the present age, such is not possible for the numerous genera (about 4,000) of fishes, but it remains possible for the families. And, yes, today's serious student aspires to know them all, even the few that are not readily recognizable.

It might interest you to learn that in recent years there is a vocal faction, arising not within ichthyology but rather herpetology and spreading into botany, that advocates rankless nomenclature—a system of names with no particular ranks attached to them. For over a century we have had in general use more ranks than the five that you used (kingdom, class, order, genus, species). I have mentioned that families have a particular significance in ichthyology. With the advent in zoological nomenclature of the family also came an immediately lower rank, the tribe. Research being what it is, intermediate ranks, such as super-, sub- and infrafamiles and tribes, are often used in specialist discussion. Rankless nomenclature would strip rank from all such arcane considerations.

Can you imagine the hapless student trying to learn ichthyology, encumbered by a jumble of names, most of which have no readily recognizable counterpart in nature? I hope that you can rest assured that ichthyology—*Ichthys'* army if I may call it that—will never surrender their families, any more than your *Flora*'s army would have surrendered their genera.

In looking through your *Systema* (first edition) I was struck by the absence of a genus for catfishes—a notable omission, for, as the late Archie Carr (1909–1987) once wrote in 1941 in *The Fishes of Alachua County, Florida: A Subjective Key*, "Any damn fool knows a catfish." The genus *Silurus* does appear in your 10th edition, and I notice that it is mentioned in all five parts of Artedi's *Ichthyologia*: in his *Bibliotheca* (part I) and *Philosophia* (II); and in his *Genera* (III), *Synonymia* (IV), and *Descriptiones* (V), but only in their Appendices not as one of his 45 numbered fish genera—an odd circumstance from a modern viewpoint. I wish you would have enlightened us about its whys and wherefores.

"Any damn fool knows a catfish"

I write from Australia, an island continent just becoming to be known to Europe in your time. There is in global use today a marvellous electronic invention called the Internet, which you, too, would have found useful in communicating with your many correspondents worldwide, even on ships at sea. For me it provides access to Artedi's posthumous publications, now available to anyone anywhere, from the Bibliothèque Nationale de France through its service called *Gallica*. If you had a computer and internet connection you could find out about it through a resource called *Google*.

I trust that you would have been startled to learn of another vocal faction—the one charging you with the intellectually unpardonable crime of essentialism, stemming either from Aristotle or Plato; to your critics it did not seem to matter which. I can imagine how you would have been taken aback—so much so that, even if it were possible for you to speak out, you might have remained silent about this sorry episode during all its years running into decades. Happily, now at an end should be this miserable history of depreciatory comment without a particle of truth. Its passing is hopefully marked by the research of Mary Winsor, particularly her paper on *The Creation of the Essentialism Story* (2006 *History and Philosophy of the Life Sciences* 28:148–174), which culminates many years of research and publication.

No longer are we routinely taught Latin in school, so the meaning of your writings has become increasingly inaccessible in their original form. We have come to depend upon one or another translation or interpretation. For me one of the most enlightening was by the French botanist, Auguste-Pyrame de Candolle, who was born within a month of your demise in 1778.

He proposed to explain the difference between *artificial* and *natural*, as they were used by early naturalists, especially you. I found his explanation particularly revealing because in the English speaking world the *artificial* was almost always seen as deficient or objectionable in some way relative to the *natural*.

Aside from you, of course, Candolle was the first, I think, clearly to point out the different purposes of artificial and natural classifications. In 1813 he wrote that these different "classifications follow different laws and rules. Nevertheless they are often mistakenly confounded. And we see in analysing them separately that all of the mistakes made in each of them stem from the attempt to introduce the principles of one into the other type of classification."

Today we do not use the term *artificial* to mean anything in particular, but everywhere students use identification keys in order to find out the name of the plant or animal in hand—provided that they do not already know it. And, yes, for plants the parts of the flower, their number and arrangement, are still routinely used for such purposes, much as you originally proposed in your sexual system of plants. Modern students are surprised to learn that in using an identification key they are in fact using, successfully and profitably, an artificial system exactly for the purpose for which it was intended.

As for the natural groups of plants and animals—the natural system— some progress has been made since your time. Everywhere today, there is much study of the hereditary material, DNA, which promised to reveal all of the natural groups that have for so long been hidden and have eluded the careful research of botanist and zoologist alike. For fishes this endeavour is still in its early stages, and so far has accomplished little except to reconfirm the naturalness of a few families of fishes—Artedi's genera in other words. In the future no doubt many more of them will be so reconfirmed.

Reconfirmed, too, will be your judgement in 1735 that ". . . the greatest Ichthyologist of our time, the Very Illustrious Dr. *Petrus Artedi*, a *Swede*, has communicated his method to us, who hardly can be equalled by anyone in distinguishing the natural genera of the fishes . . ." Or, in his words (and your citation style),

"*Ichthyologia est scientia* . . .," Art. phil. 2.

On that note of cheerful optimism, I end this letter.

Yours Sincerely,

Gareth Nelson
School of Botany
University of Melbourne
Victoria 3010
Australia

Carl Linnæus

Dear Dr. Linné,

On names: As a 'y' may be freely interchanged with an 'i' without much damage, a colleague once asked me if I was named after you. My parents surely did not know of any Linné, or Linnaeus for that matter. Household names for me growing up in mid-20th century New York City were Lucille Ball, Frank Sinatra, Yogi Berra, Beverly Sills, Adlai Stevenson, the popular entertainers, athletes and politicians of the era. One name, then as now, could readily identify a celebrity, but it was 'Lucy' not 'Linné.' Don't take it personally, but scientists tend not to be such household names. Maybe Copernicus, Galileo, Newton and Darwin, who came after you, but not Linnaeus.

My mother thought carefully about names, though. For her, names should be easy to spell and pronounce. Her three daughters are named simply Jane, Diane and Lynne.

Names of plants and animals were equally important, but it was mainly the common names that I learned for the plants in our garden or the animals in the woods or along the shore. 'Poisonous yew bushes', yes; but '*Taxus baccata* L.'? Who knew?

In Ichthyology class, as I was introduced to a classification of fishes, I listened carefully to my professor and dutifully wrote down facts about a fish family named the Harrisons. At least, that's what I thought he said. So foreign were even vernacular names reflecting a scientific classification that I misheard 'Harrisons' for 'Characins,' short for the Characinidae or Characidae, the tetras and their relatives.

In the 12th edition of the *Systema Naturae*, you published names of fishes—with a little help from a fellow student at Uppsala, Peter Artedi—that will follow me all my days. As a student, I learned quickly to appreciate the need for a universal scientific name for fish species. *Pomatomus saltatrix* (Linnaeus, 1766) is Bluefish, unless you are in Australia where it is Tailor, or the east coast of South Africa where it is Shad, which should not be confused with American Shad, *Alosa sapidissima* (Wilson, 1811)—not one of yours—known also as sábalo americano (Spanish) or alose savoureuse (French), among many colloquial names[1].

Your system of a binominal was sanctified. *Pomatomus saltatrix* (Linnaeus, 1766), along with a detailed scientific description, is one encyclopedic entry in the vast catalog or classification of life.

On places: Species names reflecting distribution were meant to link a taxon with the region in which it lives, but, without knowing precise distributional limits, the practice was flawed: the flightless Greater

The Bandwing Flyingfish *Cheilopogon exsiliens* (L.) (Exocoetidae) was described by Linnaeus in 1771 in a book mostly devoted to plants (*Mantissa Plantarum*)

Rhea, *Rhea americana* (Linnaeus, 1758), for example, lives not throughout the Americas, but solely in lowland habitats in southeastern South America. If only you had known.

Knowing the place where an organism lives is as critical a piece of information as are details of when it reproduces or what it eats. The concept of endemism is universal and as old as the study of natural history: a species is endemic to an area if it lives there and nowhere else. As I began to study biogeography along with systematic ichthyology, I became aware that places too need universal names and descriptions so that they can be readily understood by all and placed in a classification. Taxonomists describe taxa and place them in a classification; biogeographers describe natural areas and place them in a classification. The parallels between systematics and biogeography became obvious: taxa are to systematics what areas are to biogeography. Moreover, the classification of taxa and areas are inextricably linked.

Rules for naming areas, analogous to those for naming taxa, have been formulated in the International Code of Area Nomenclature (ICAN) by the Systematic and Evolutionary Biogeographical Association (SEBA)[2]. The ICAN is a first step in laying out ground rules for the recognition, description and classification of natural areas. It does not endorse any particular systematic or biogeographic philosophy; its goal is for scientists to give names of places the same attention they give names of taxa.

Exploration and discovery, combined with advances in systematic methods and applications, have meant continuous revision and updating of biological classification. The same cannot be said for biogeographical classification which has largely remained stable for over a century. How can this be?

The 19th century ornithologist Philip Lutley Sclater aimed to uncover what he considered the ontological or essential divisions of the earth's surface as reflected by organic distributions. In 1858, he proposed a division of the world into six biogeographic regions: Nearctica, Paelearctica, Neotropica, Æthiopica, Indica and Australiana[3]. These were readily adopted by Alfred Russel Wallace and contemporaries. By 1894, Wallace deemed Sclater's regions ". . . old-established and widely accepted[4]." Never mind that they did not address marine regions or trans-oceanic terrestrial distributions, they were considered sufficient for describing the distribution of taxa and are used to this day.

What of area relationships, a classification? Endemic biotic areas, defined by the taxa that live in them and nowhere else and delimited by taxic distributional boundaries, are what biogeographers discover and describe. Relationships among the endemic biotic areas form the basis of area classifications. Sclater's Neotropical region was divided by Wallace into four subregions: Chilean, Brazilian, Mexican, and Antillean[5]. But, systematic studies across a range of taxa have revealed the following homology: the Chilean biota is more closely allied with the southeastern Australian biota than either is to the Mexican biota. That is, the Neotropical Region is unnatural; it does not comprise a coherent biota. Keeping it in a classification of areas is analogous to keeping the Reptilia in a classification of tetrapods. It works in the sense that any classification does, but we can do better.

Area classifications should have high information and predictive value and can do so only if they are linked to taxonomic classifications that have as a goal the recognition of natural (monophyletic) groups. This may not have been the way you would have said it, had you been a biogeographer. But, I share your goal: to understand the natural order of life on earth. And, all in my household now know the name "Linnaeus."

Lynne R. Parenti

Curator of Fishes and Research Scientist
Department of Vertebrate Zoology, MRC NHB 159
PO Box 37012
National Museum of Natural History
Smithsonian Institution
Washington, D.C. 20013–7012

Notes:

[1] Nelson, J.S., Crossman, E.J., Espinosa-Pérez, H., Findley, L.T., Gilbert, C.R., Lea, R. N. & Williams, J.D. 2004. *Common and Scientific Names of Fishes from the United States, Canada, and Mexico*. 6th Ed. American Fisheries Society Special Publication 29, Bethesda, Maryland.
[2] Ebach, M.C., Morrone, J.J., Parenti, L.R. & Viloria, A.L. 2008. International Code of Area Nomenclature. *Journal of Biogeography* 35, 1153–1157.
[3] Sclater, P.L. 1858. On the general geographical distribution of the members of the class Aves. *Journal of the Proceedings of the Linnean Society: Zoology* 2, 130–145.
[4] Wallace, A.R. 1894. What are zoological regions? *Nature* 49(1278), 610–613.
[5] Wallace, A.R. 1876. *The Geographical Distribution of Animals; With a Study of the Relations of Living and Extinct Faunas as Elucidating the Past Changes of the Earth's Surface*, Macmillan, London. 2 vols.

These specimens of the Butterfish *Peprilus alepidotus* (L.) (Stromateidae) were sent to Linnaeus from the Carolinas in 1763 as "Angel Fish"

Carl Linnæus

Dear Carl,

Wow! I've just picked up your *Species Plantarum* and have had to go for a lie down. Brilliant, it took my breath away: all known plants in one book. OK, you might have missed a few, but I think you've shown us the way forward with your vision. As mere mortals we need to distil the complexity of life into a system that we can begin to comprehend. We will discover more and more about individual species, but unless we can understand that knowledge in the context of life on our planet and communicate it widely, we are little more than misers, hoarding away gold in dusty pots. You've given us the framework to invest our knowledge and build systems which will allow us to better manage our natural resources and our relationships with the species that share our planet.

By the way, I think your idea of a binomial system of nomenclature is utterly cool. Short, simple

Linnaeus' own copy of *Species Plantarum* (1753) in the Linnean Society of London

and easily understood. Having this system of readily communicable, unique labels for each species greatly enhances our ability to share knowledge and to better understand the world around us.

As I sit here, we still lack a widely available working list of all known plant species. In 2002 this was called for by the 190 countries who signed the Convention on Biological Diversity. The politicians recognise the importance of having a working list of organisms in order for us to manage our information to better conserve species and to interact with them in sustainable ways. Some think that it will be impossible to get experts to agree and a list will never be produced. Others think that such a list already exists and are surprised when they realise that something as basic and as necessary as this does not exist already. I take solace from your achievements and realise that with dedication and vision such a list can and must be produced. I have great hopes that a

Lamium album L. (Lamiaceae), or the White Deadnettle, of this correspondent's study group

list, at least covering plants, will be available by 2010, facilitated by the world wide botanical community working together.

I guess you don't get out much now, but if you did, I think you'd still see the impact of your work. Initiatives such as the Global Biodiversity Information Facility, the Catalogue of Life, and the Encyclopedia of Life aspire to your vision of comprehensive coverage of life on earth. I hope that the systems we build now can inspire future generations as you have inspired us.

Best wishes,

Alan Paton

Dear Baron Linné:

I'm pleased to have this opportunity to add my congratulations to your anniversary celebration—250 years is quite a run! Since we haven't met, a brief personal history may be appropriate. I'm an arachnologist, having spent almost 40 years now working on the systematics of spiders and ricinuleids (though I'm hardly ancient—I just got an early start!). The Ricinulei, incidentally, are an unusual order of arachnids that may be unfamiliar to you, as the first known species (a fossil that was initially misidentified as a beetle!) wasn't unearthed until 1837, and the first recent species wasn't described until a year later.

My professional career has been spent mostly at the American Museum of Natural History in New York, but my doctoral work was done in the green pastures of Harvard University, where our catechism was supplied by Ernst Mayr and his 'New Systematics' (or, as I prefer to call it, the 'Non Systematics'). In retrospect, I should have known that something was amiss from day one. There I was, unpacking my stuff into the tiny cubicle I was assigned on the fourth floor of Louis Agassiz's Museum of Comparative Zoology, when a distinguished-looking gentleman strolled in, and asked a few questions about who I was, what I was interested in, what my thesis was going to be about, and the like. I was scheduled to take an evolution course that first semester, which was to be team-taught by Ernst Mayr, erstwhile malacologist Steve Gould, and lepidopterist John Burns. The assigned textbook, of course, was to be Mayr's *Animal Species and Evolution*, so I had a copy of it out on my desk. My visitor pointed to the book and said "Oh, I see you've got a copy of the bible." That seemed like a very strange comment, indeed; hailing, as I did, from the fundamentalist bible-belt of the Appalachian Mountains, I wasn't exactly used to hearing evolutionary texts referred to as biblical, in any sense! So I just shrugged the comment off, and went on with the conversation. After a while, my guest stood up, shook hands, said "Oh, by the way, I'm Ernst Mayr", and strode out.

Be that as it may, I was soon properly indoctrinated, but I suspect you'll find the details of that mantra somewhat odd. Species, I was told, are real, and what every right-thinking organismic biologist should focus on, to the exclusion of all else, because groups of species—higher taxa—are not real, but merely human and artificial constructs.

Now, I have to say that I never found that position to be very sensible. I could look around at my major professor, for example, an arachnologist by the name of Herbert Levi. His mentor had been the long-time spider curator at the American Museum, a true seat-of-the-pants taxonomist named Willis Gertsch. By this time, however, on at least a few occasions, Levi had published a revision of a group, and Gertsch had felt compelled to write a rebuttal. Levi then had to publish papers attempting to explain why some of his species were, at least according to Gertsch and other highly respected colleagues, just artificial assemblages of several different and easily distinguished species that weren't even necessarily each other's closest relatives. Yet, despite all these disputes about species, none of these eminent systematists seemed to have any difficulty whatever in agreeing on at least some higher taxa, such as spiders.

Indeed, spiders seemed, then and now, to be the quintessential example of a natural, or real, group—so real that I thought it was utterly impossible for any of your intellectual descendants to make a mistake about whether a given organism is or is not a spider. Now, I do have to stop and be honest here; what I actually thought was utterly impossible was for anyone other than a paleontologist to make that mistake. Alas, I was wrong; a few years back I had to co-author a paper on *Brucharachne*, the spider that wasn't—a spider, that is. In 1925, as it transpires, one of the worst taxonomists South America ever produced, the immortal Candido Firmino de Mello-Leitão, actually described a whole new family of spiders, based on what turned out to be a single male mite.

But I bring all this history up mostly to point out that we have now made our way fully to the opposite pole on the rather fundamental issue of taxa and their reality, with papers appearing in the journal *Systematic Biology* about so-called "phylogenetic taxonomy" and bidding "a farewell to species." In other words, some of our colleagues are now arguing exactly the opposite point of view from the 'non systematics' I was taught as a grad student; they are arguing that higher taxa are real but species, somehow, are not. At least one author, Fred Pleijel, adopts "a view where species simply are denied any role in taxonomy and where only monophyletic groups are recognized by formal Latin names" and he says he applies "uninomials for all names, because this treatment does not recognize any species entities and applies the same nomenclature for all taxa."

As you might guess, I personally consider both of these polar positions to be preposterous, wrong-headed, and insidious. Let's proceed by getting rid of the obvious. I assume we both recognize that species, and groups of species, are first of all hypotheses, and in that sense only, represent artificial constructs of the human mind. But I assume we both recognize that it is possible, at least in theory, for one of those hypotheses to be an accurate statement about the real world. Who knows, on a good day, maybe even more than one of our hypotheses might be accurate!

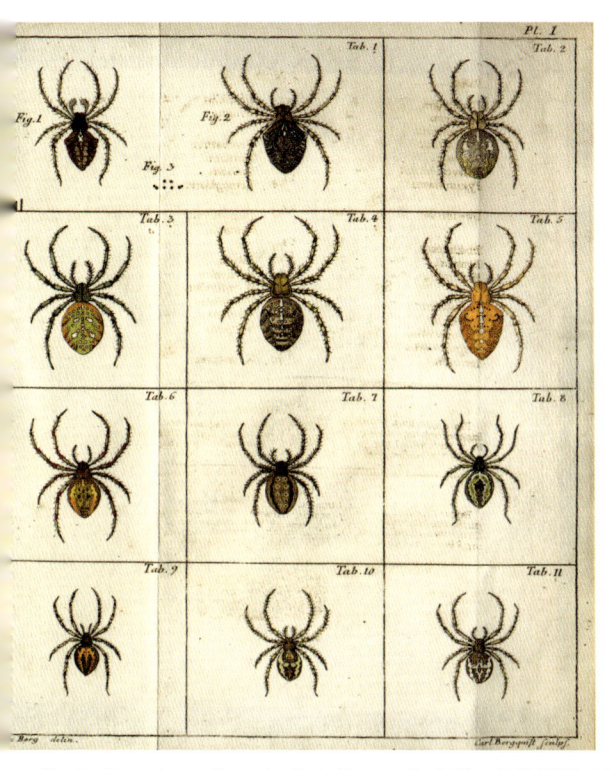

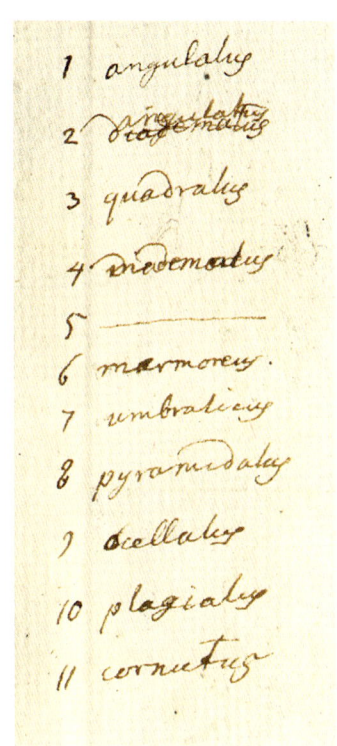

Illustrations from *Svenska Spindlar* by Carl Clerck (1757). Linnaeus' copy is heavily annotated.

Throughout the centuries since Clerck (and here I must beg your forgiveness, for spider systematics, in its typically precocious fashion, starts with your countryman Carl Clerck's *Svenska Spindlar*, a fine book published in 1757 and no doubt familiar to you, since it preceded the 10th edition of your classic *Systema Naturae* that serves as the starting point for the remainder of zoological nomenclature), most systematists have avowedly been seeking natural classifications, by which they meant classifications that consist of species, and groups of them, that actually exist in the real world.

And I would suggest that this is an entirely appropriate goal. After all, how do we know that anything, such as the chair I'm currently warming, actually exists? The answer, of course, is that we can keep bumping into it—lots of different sources of evidence agree. So let's look at a putatively natural group like spiders, and ask whether we have evidence that it is real.

Spiders have at least two characters that are completely universal within the group, and completely unknown outside the group. The most obvious character is the abdominal spinnerets through which silk is emitted. But there is another unique character, namely that adult males have modified structures on the tips of their pedipalps that are used to transfer sperm to the females during copulation. In a male spider, there is no anatomical connection between the gonads and the intromittent organ; the male spider has to deposit a drop of sperm from his abdomen, and then dip his palps into the sperm to charge them for mating.

To date, systematists have described some 1.75 million species, and of those, some 40,000 species are currently considered valid members of the Order Araneae, the spiders. If we pick abdominal spinnerets as a character, that feature allows us to group 40,000 species, and no others, as spiders. If we then group together those species with male pedipalps modified for sperm transfer, we are once again grouping together a mere 40,000 out of 1.75 million species. In statistical terms, imagine that you were asked to draw 40,000 species from a pool of 1.75 million, and then to repeat that procedure. What would be the probability of picking exactly the same set of 40,000 species, the second time, by chance alone? In fact, that probability is so infinitesimally small that in this case no real statistician would need Joe Felsenstein's preferred three characters to conclude that spiders, as a group, are one of the best corroborated hypotheses around. When you consider in addition the combinability of those two characters with countless others supporting groups both larger (arachnids, arthropods, etc.) and smaller (e.g., the families, subfamilies, genera, etc. of spiders), the evidence becomes overwhelming—each of those characters could potentially contradict the grouping of 40,000 species as spiders, but they don't.

So I suggest that at least some of the groups in your classifications, and mine, are real, natural, or monophyletic—choose whichever adjective you prefer, since the choice actually makes no difference whatever to anything important. No one would claim that all our groups (in your time or today) are real, but surely we would both like our classifications to consist entirely of groups like spiders—groups based on the congruence and combinability of different sources of evidence.

By the same token, what is now called the Linnaean hierarchy—your system of recognizing groups within groups—is flexible enough to allow systematists to name all the groups they care to hypothesize and talk about, and to reflect, in as much or as little detail as they choose, the hypothesized phylogenetic relationships among those groups. Indeed, your hierarchy of groups within groups provided the pattern that evolutionary theory seeks to explain, and it performed as admirably and effectively in Darwin's day as it did in your own time, and as it continues to perform today.

It seems very strange, therefore, that some systematists have recently begun to argue that we need to abandon the Linnaean hierarchy entirely. You can often recognize these folks easily enough; they're the ones wandering around in t-shirts that read "Phyla Schmyla" (which is apparently a lapsus or mere printer's error; the shirts were actually supposed to have read "Yale Schmale"). And, of course, these are the same folks now promoting the so-called PhyloCode as the cure to all our woes.

In my view, though, the PhyloCode asks far too little of biological nomenclature, and there is no compelling reason whatever to set our sights that low. Let me explain first by quoting from the PhyloCode's architects: "an underlying principle of the PhyloCode is that the primary purpose of a taxon name is to provide a means of referring unambiguously to a taxon." That's certainly fine, as far as it goes, but they actually go much farther, adding (and I emphasize): "not to indicate its relationships."

Now, on the surface, that seems like just a totally absurd viewpoint. If taxon names *can* convey information about relationships, why on earth would any systematist prefer that they *not* do so? Well, PhyloCoders wring their hands because including information about relationships can result in name changes when hypotheses about relationships change. In their view, that loss of stability "is too high a price to pay for incorporating taxonomic information . . . into the names."

I wonder whether such an argument could actually even be presented seriously in any other branch of science. What is the scientific justification for regarding stability as more important than, say, accuracy, in any context? I'm not aware of anything in the philosophy or methodology of science in general that could justify the claim that stability is more important than, say, information content. The entire idea of science is to overthrow hypotheses when they fail to meet the test, to replace them with improved hypotheses that perform better, and to do so as quickly as possible! Surely not even a PhyloCoder would argue that we should stick with the poorer hypothesis so as to achieve stability of anything whatsoever!

In their case, however, the argument is even weaker; the PhyloCode may ask too little of nomenclature, but it actually delivers even less, because PhyloCoders are not arguing for keeping the meaning of names stable (i.e., allowing no changes in the taxa they contain), but merely for keeping the spelling of names stable, even in the face of dramatically changing taxonomic content. So their proposition is actually quite stark: according to them, we should abandon completely the goal of including information about relationships in taxon names in order to achieve the supposedly more important goal of stability (in spelling).

These days, the PhyloCoders seem willing to abandon just about everything else as well. It seems that there were at least two motivations behind their movement, which I'm ashamed to say was originated by two herpetologists from the United States, Kevin De Queiroz and Jacques Gauthier. One motivation, seemingly, was just to find some way to salvage non-monophyletic groups like Reptilia and Dinosauria, so that herpetologists would not be inconvenienced by increased understanding of amniote interrelationships. But the main motivation, and I'll quote from De Queiroz & Gauthier, is that:

> if the Darwinian Revolution is ever to occur in biological taxonomy . . . then the role of the principle of descent must change . . . from an after-the-fact interpretation to a central tenet from which the principles and methods of taxonomy are deduced. . . . Previously, taxa were considered to be defined by characters and only interpreted after-the-fact as products of evolution.

In the benighted view of these authors, most if not all previous systematists have been ignorant or misled essentialists or typologists who have been so stupid as to use character-based definitions or diagnoses of taxa. Now one thing I've learned from 40 years of watching biologists is that whenever you find one systematist calling another one an essentialist or a typologist, you can be 100% sure that the name-caller is purely, simply, and entirely, wrong, and is just creating a smokescreen to cover his or her tracks. In this case, there are a lot of tracks to cover, and they rank about as high on the SIP scale as any I've encountered; the SIP scale, incidentally, measures the levels of Sanctimony, Inflation, and Pomposity. Rest assured, these authors truly believe that their proposals have already created (and here again I quote), a "new era in biological taxonomy."

The complaints of these workers about the non-evolutionary basis of the existing system are entirely specious and unconvincing exercises in metaphysics, and are not worthy of serious attention. Let's look instead at what they suggest is an improved system, which they argue will do a better job of promoting explicitness, universality, and stability of names, as now encapsulated in the draft PhyloCode. To accomplish this goal, they wish to define the names of taxa in terms of common ancestry, rather than characters. My colleagues Kevin Nixon and Jim Carpenter have therefore referred to this view as the Node-Pointing or NP system. Now that's a highly unfortunate choice of terms, since NP is a fine pair of initials that doesn't deserve to be sullied in such fashion (even my wife's initials are NP, which I guess makes us an NP-complete family). I'll call it the NB (or Node-Based) system instead, since that is the phrase actually used by de Queiroz and Gauthier themselves, and "N.B." has all the appropriate connotations.

The NB system was originally depicted as the true and long-delayed culmination of the Darwinian revolution in systematics, a promised land from which those awful typologists would be banished and in which only their own, truly Darwinian, properly ancestor-worshipping groups would be allowed in classifications. Nowadays, though, we're told that, never mind, their system can accommodate paraphyletic groups just as easily as can the current one. And although they originally argued that it is the use of ranks that generates the intolerable instability of spelling in the current system, and that ranks therefore have to be trashed like yesterday's newspaper, nowadays we're told, never mind, you can have your ranks and PhyloCode too (at least so long as there is no connection between ranks and the spelling of names). I guess these folks haven't figured out that trying to be all things to all people is the quickest way to become nothing to anyone.

In any case, those authors and their fans have repeatedly illustrated:

Three classes of phylogenetic definitions
(a) A node-based definition is used to define the name of a clade stemming from the most recent common ancestor of two specified organisms, species or clades (e.g., Aves = the clade stemming from the most recent common ancestor of *Struthio camelus* and *Passer domesticus*).

Of course, you and I both realize that De Queiroz and Gauthier don't actually know anything at all about the most recent (or any other) common ancestor of those two species; like us, they actually know only the characters that happen to optimize at the node of their preferred cladogram that happens to subtend those two taxa.

(b) A stem-based definition is used to define the name of a clade of all species sharing a more recent common ancestor with one specific organism, species or clade than with another (e.g., Lepidosauromorpha = Lepidosauria and all species sharing a more recent common ancestor with Lepidosauria than with Archosauria).

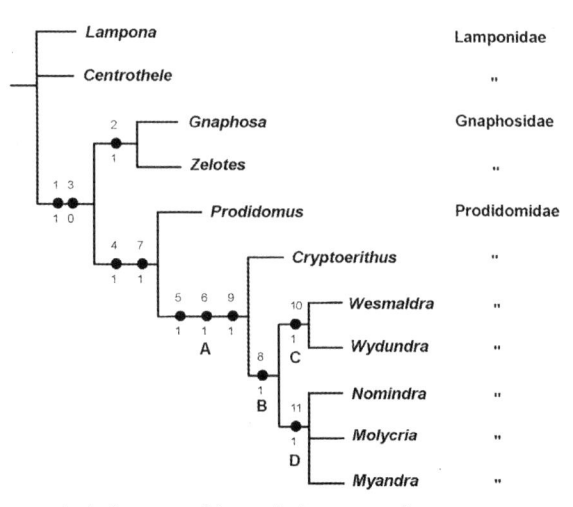

A cladogram of Australasian ground spiders (see Platnick & Baehr 2006. *Bulletin of the American Museum of Natural History* 298: 1–287 for explanation)

(c) An apomorphy-based definition is used to define the name of a clade stemming from the first ancestor to evolve a specified character (e.g., Tetrapodomorpha = the clade stemming from the first vertebrate to evolve pentadactyl limbs).

Of course, these are all old ideas, thoroughly discussed by Willi Hennig and others in the context of fossil fragments, stem groups, crown groups, and the like. A favorite example of these authors is the lizard family Agamidae, and here again, I'll quote:

> For example, the name 'Agamidae' might be defined as the clade stemming from the most-recent common ancestor of *Agama* and *Leiolepis*. Such a definition is thoroughly evolutionary in that the concept of common ancestry is fundamental to the meaning of the name.

As if the definition could possibly be one wit less evolutionary if it concerned itself with the evidence from which those author's conclusions about common ancestry were drawn (after-the-fact)!

The reason they like this example is that the Agamidae is another one of those groups, like Invertebrata, Reptilia, and Dinosauria, that may be artificial (i.e., non-monophyletic). Although these largely Old World lizards were formerly divided into two separate families, the Agamidae and Chamaeleonidae, some agamids may actually be closer to chamaeleons than to other agamids. To these authors, this is apparently a tragedy. They attribute this insight to some of their herpetological mentors publishing in the late 1980s, but a little more concern with history might have been appropriate. Back around 1936, a German zoologist by the name of Willi Hennig was publishing papers on lizards, and if you look in the books he later wrote about something he called phylogenetic systematics, you'll find that (gasp) agamids are traditionally recognized are paraphyletic because the chamaeleons are excluded from the group.

Insights such as this one prompted Hennig to develop phylogenetic systematics, not so-called phylogenetic nomenclature! He didn't seem to regard it as any great tragedy that the agamid/chamaeleonid dichotomy may need to be expunged from our classification, because it denotes a hypothesis of relationships that may be inaccurate.

Like any other reasonable systematist, he would have asked first, where on the cladogram the type genus of the Agamidae happens to fall, and he would then have chosen a solution that conveys the phylogenetic information accurately while doing the least possible damage to existing concepts. Probably the easiest solution would be to restrict Agamidae to the group including *Agama* and supply a new family-group name for the remaining taxa previously misplaced in the Agamidae (the leiolepids). Alternatively, all three clades could be lumped into a single family, a solution de Queiroz and Gauthier dislike because the name Chamaeleonidae happens to have priority over Agamidae.

Instead, of course, those authors argue that we need to adopt a system in which the dreaded but perfectly acceptable name Chamaeleonidae retains its association with those taxa, the name Agamidae is retained to refer to a more basal node, and we just abandon the ranks so that "Chamaeleonidae is now judged to be nested within Agamidae" and Agamidae now actually becomes synonymous with the older group name Acrodonta, which happens to be above the family-group level in rank and is hence unregulated by the current zoological code.

But ultimately, the question is what, exactly, it is that remains stable. In their lizard example, they suggest that:

> Under phylogenetic definitions, 'Chamaeleonidae' retains its association with the clade stemming from the most recent common ancestor of the species represented by filled circles, and 'Agamidae' retains its association with the clade stemming from the most recent common ancestor of the species represented by open circles, although the chamaeleonid species are now also thought to have descended from this ancestor. The manner in which the definitions are stated ensures that no names designate paraphyletic taxa, and neither splitting nor lumping occurs, but hierarchical relationships may be altered (e.g., Chamaeleonidae is now judged to be nested within Agamidae).

As the last comment shows, what has actually been achieved, by abandoning Linnaean ranks and categories, as they advocate, is merely stability of spelling. Here is the Linnaean classification before the altered concepts of relationship:

> Family Agamidae, containing taxa ABCDEF
> Family Chamaeleonidae, containing taxa GHI

This Linnaean classification accurately reflects the traditional view of the interrelationships of these taxa, under which taxa A, B, and C form the sister group of D, E, and F, with taxa G, H, and I representing the sister group of A–F together. What matters, here, however, is not that particular cladogram. What matters is that from the information in these two lines:

> Family Agamidae, containing taxa ABCDEF
> Family Chamaeleonidae, containing taxa GHI

and that information alone, any systematist can deduce a whole series of three-taxon statements of relationship, all of which must be true if the classification is true: (AB)G, (AB)H, (AB)I, (AC)G, etc.—if I've counted correctly, there are actually 63 such three-taxon statements that follow from this classification. Such inferences are possible solely because of the use of Linnaean categories: the classification asserts that taxa A-F are not chamaeleonids, and that taxa G-I are not agamids, and it is precisely those prohibitions that allow the detailed hypotheses about relationships. As Karl Popper has shown, it is also precisely by making prohibitions that hypotheses become testable and hence scientific.

Contrast that with the de Queiroz and Gauthier solution, which looks like this:

> [unranked group] Agamidae, containing taxa ABCDEFGHI
> [unranked subgroup] Chamaeleonidae, containing taxa GHI

From this classification, one can deduce only 18 three-taxon statements: (GH)A, (GH)B, etc. They have indeed kept the spelling stable, but at the cost of reducing the information content of the classification by about 75%! And the name Agamidae, although still spelled the same, now refers to a different group of taxa, all the animals previously placed in the Agamidae PLUS all the animals previously placed in the Chamaeleonidae. And this stability would be institutionalized so that the PhyloCoder's usage would become immortal. If tomorrow we discover that birds are more closely related to chamaeleons than to crocodiles or any other living taxa, then Aves would also become a member of the Agamidae. It would be hard to imagine a case that would better fit Gene Gaffney's quip that stability equals ignorance.

According to de Queiroz and Gauthier, "a name should not designate different taxa [I guess they mean, unless the name is Agamidae], nor a taxon be designated by different names, at different times." By that standard, it is difficult to see any benefit whatever to node-based nomenclature. Of the two original group names, one now designates different taxa, and information content has been strangled. If instead we opt for:

> Agamidae, containing taxa ABC
> DEFidae, containing DEF
> Chamaeleonidae, containing GHI

(i.e., a conventional Linnaean classification, albeit one implying nothing more than a basal trichotomy among these three families), again one name now designates different taxa, but information content decreases only to 54 three-taxon statements, rather than 18.

If instead we opt for the more resolved, fully subordinated classification:

>   Chamaeleonoidea, containing taxa ABCDEFGHI
>   Agamidae, containing taxa ABC
>   Chamaeleonidae, containing taxa DEFGHI
>   DEFinae, containing DEF
>   Chamaeleoninae, containing GHI

Two names now designate different taxa (although one of them also remains accurate, with just a slight change in spelling) but information content zooms to 81 implied three-taxon statements.

If you are unhappy with implied three-taxon statements as a measure of information content, then try other measures; based on the comparisons done by Mary Mickevich and I, the results are unlikely to differ significantly. But consider for a moment what classification is all about. Is stability the primary goal of classification? Of course not! The primary goal of classification is and always has been maximal predictive power. Cladists have long argued that phylogenetic classifications are both the best summaries of the limited character information already available, and the best basis for making predictions about the much larger universe of as yet unstudied characters. Taxonomists are in the business of providing highly predictive classifications on the basis of extremely small amounts of data, and our track record of success, from your days on, is none too shabby. Every three-taxon statement that a classification implies represents a prediction that whatever future synapomorphies might be found, in any character system whatever, may fit the pattern (A & B as opposed to)C but will not fit the two conflicting patterns (A & C as opposed to)B or (B & C as opposed to)A. Again, scientific classifications prohibit things.

In contrast, the NB system achieves stability in spelling only, and often at great expense in terms of sacrificed information content and predictive power. Ultimately, no NB system can possibly produce a more predictive classification than would a fully informative Linnaean hierarchy based on the same cladogram. In other words, there is no possible potential gain in predictive power to be achieved by switching to an NB system, and many possible potential losses of that power.

Indeed, if unlike the PhyloCoders, one set out with the goal of purposefully including the maximal possible amount of taxonomic information in group names, it would be hard to design a system that is more efficient than yours—that could pack more information into a single taxon name than the current hierarchy does for generic and family-group names in zoology (and even ordinal names in botany). One way of doing that, at relatively small cost in my view, would be to extend the use of standardized endings to taxa above the family-group level (e.g., to the orders, classes, and phyla of animals). Because they're unregulated by the current zoological code, those names aren't particularly stable at the moment anyway; for example, the order I work on is called Araneae by most folks, but workers from the former Soviet Union use Aranei instead. Changing to standardized endings would be pretty simple, would eliminate those kinds of silly disparities, and would extend the immediately apparent exclusivity in group membership that makes generic and family-group names so very informative and useful today.

Under the PhyloCode, though, all such information is purposefully discarded, so that identifying a specimen as a member of any named group no longer tells you anything whatsoever about the relationships of that specimen. Under the PhyloCode, a member of the Agamidae might actually be more closely related to a member of the Chamaeleonidae than to another agamid—in other words, the original problem with the name Agamidae still persists, but we've added the wrinkle that our member of the Agamidae might also be more closely related to a member of the Drosophilidae, or the Hominidae, than to another agamid, in so far as you could ever tell from its name. Now that's real progress!

The Hardhead Sea Catfish *Ariopsis felis* (L.) (Ariidae) is found in the western Atlantic and Gulf of Mexico

Let me close with a simple example of this difference involving (of course) spiders. I was wandering around John Murphy's garden out in Hampton, England, and came across a nice jumping spider. Now, jumping spiders, the family Salticidae, are probably the easiest of all spider families to recognize. With their large anterior median eyes, their excellent vision, the often highly exuberant and ornamented morphology that males use in their elaborate courtship displays, and their prowess at jumping on prey several body-lengths away, salticids are quite distinctive. Probably the vertebrate analog would be best exemplified in Archie Carr's *Subjective Key to the Fishes of Alachua County, Florida*, in which the first couplet reads "Any damned fool knows a catfish." The botanical analog might be the mints, which may be the only plant family duffers like me can manage to recognize successfully, wherever we roam!

A fair chunk of my time over the past 20 years has been spent on The World Spider Catalog, which is now easily accessible via the aptly-named World Wide Web. The catalog provides a listing of all the currently valid spider species, all the other names that have ever been applied to them in the past, and all the citations to every important taxonomic treatment ever published on every one of those species, from your time to the present. To my knowledge, the only thing like it currently available for any other sizable group of organisms is Bill Eschmeyer's catalog of the fishes. If you visit the spider catalog, you'll find a summary table that shows, for each of the 108 currently recognized spider families, the numbers of currently valid genera and species, including, at the very end of the list, the salticids, with 5,088 species.

Using the Linnaean hierarchy, when I identified the spider in John's garden as a salticid, I was asserting that John's spider is more closely related to any single species currently included within the Salticidae than it is to any single species that is currently excluded from that family. In other words, if my identification, and the current classification, are both correct, then John's spider is more closely related to salticid species #1 than it is to any of the 34,794 spider species currently excluded from the Salticidae. It is also more closely related to salticid species #2 than it is to any non-salticid spider. So, assuming that the spider from John's garden belongs to one of the currently

This salticid spider from California is more closely related to other salticids than to all other spiders by virtue of its shared, derived features

known 5,088 salticid species (certainly a fair assumption, at least for a British spider), then my identification enables 5,087 (other salticids) times 34,794 (non-salticids) three-taxon statements. So by placing the animal as a salticid, the current Linnaean hierarchy allows me to make 176,997,078 three-taxon statements about it, within spiders alone. If I were to expand the arena to include all arthropods, or all life, the number of implied three-taxon statements would, for all practical purposes, approach one-third of infinity—the other two-thirds would be prohibited. That's none too shabby, for a single word—Salticidae (admittedly, in a context provided—solely—by the Linnaean hierarchy and the mutual exclusivity of equally ranked names it requires).

Contrast that with an identical list of names that happen to end in 'idae,' and the numbers of genera and species they contain, but now under the assumption that the NB system is in use. Salticidae still refers to a group, that is still presumably monophyletic, but one can no longer infer anything about the status of the species included in other groups. Thus, for example, even the first family on that list, the Liphistiidae, could in fact, in the current classification, be just a subgroup of the Salticidae. The fact that both names end in 'idae'—in the NB system—does not prevent either group from being a subgroup of the other. Now, in fact, calling John's spider a salticid does not allow me to make even a single three-taxon statement involving any other species. I can't say that it must be more closely related to another salticid than to a liphistiid, because Liphistiidae might in fact constitute the sister-taxon of John's species, for all I know (or, more accurately, for all the names in an NB system could ever let me know; unlike an NB systematist, I do actually know better than that!).

So, my dear Baron, it seems that we have John's salticid and either your hierarchy, with 177 million predictions about spiders alone, or the NB system, with no useful predictions at all. The Linnaean system, with built-in exclusivity that forces classifications to prohibit things and thereby become scientific hypotheses, or the NB system, in which classifications prohibit nothing and names, by themselves, become mere propaganda, like the PhyloCode itself. My vote's with you!

Cheers,

Norm Platnick

Polaszek 185

Rue Linné 5
75005 Paris, France
26 August 2008

Esteemed Professor von Linné,

This year 2008 marks 250 years—a quarter of a millennium—since your 10th edition of *Systema Naturae*, published in 1758. Zoologists have adopted that edition, with a designated publication date of January 1st, as the starting point for all animal nomenclature. Today, in Paris, zoologists have congregated from all around the world to celebrate this occasion and to pay you the respect and homage that you so deserve. During the past 250 years the number of animals described has increased from your 4,397 to approximately 1.8 million. The task of cataloguing these species, a task that you initiated, is now on a monumental scale; it is highly complex and requires a worldwide team of people working in collaboration. Fortunately the art and science of writing and recording have progressed greatly. We now have machines, each of which can perform the tasks of a million human brains, and there are tens of millions of such machines. They have memories to match, and furthermore we are able to use them to communicate with all parts of the inhabited globe, usually within seconds.

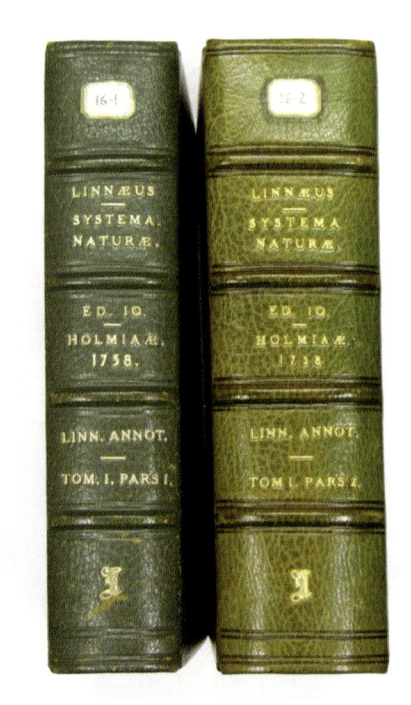

Linnaeus' own two-volume copy of *Systema Naturae, edition 10* (1758) in the Linnean Society of London

Reading this news you might think that the completion of your enterprise, as one of our most esteemed zoologists has referred to the cataloguing of the Earth's species, might have been completed long ago, but this is far from being the case. In fact (to the astonishment of many of us), we have not only failed to catalogue the 1.8 million life forms so far discovered (not even all the genera), but we now estimate that the true number (including as yet undiscovered species) is at least an order of magnitude greater. Why have we been such abject failures? You may find it difficult to comprehend that the world's population has risen from 800 million during your time, to more than 6,000 million 250 years later. This single fact has had consequences on almost every aspect of human activity, including our ability (or inability) to document the species with which we share this planet.

The sad truth is that humanity is perpetually involved in a global struggle against itself. There is war, famine and poverty on terrifying scales, and it is partly for this reason that the perceptions of many have shifted away from the encouragement of art, science and discovery. While advances in food production methods have reduced hunger in many parts of the world, and medical innovations have greatly alleviated suffering and prolonged life, one of the consequences has been this huge population growth that threatens the very survival of the planet.

So priorities have changed, and in fact many otherwise highly educated people now question the need for an effort to record the world's living and extinct species. They are mistaken, for as you showed so long ago, humanity cannot progress without knowledge of the world's living organisms. We now know that all these organisms, and probably all life that has ever existed on Earth, are related by common descent with modification; that life evolved rather than having been created by a deity.

It is possible that some of this may shock you, but perhaps coming to terms with the fact that humanity alone is responsible for its wellbeing and survival surely engenders a sense of responsibility that is lacking when all can otherwise be apportioned to a god or a devil. There is a real sense of magic in the discovery of new species—species that are either new for a particular fauna or that are completely new to science. I have experienced this excitement on many occasions, whether discovering already known insect species that are new to England or Europe, or genera completely new to science from the rainforests of the tropics. Our knowledge that all life is related, and the tools and techniques we now have that enable us to unravel these relationships, give to that magic of discovery a further, even more thrilling, dimension.

The complex habitat of the tropical rainforest understory in Vietnam

We now have the resources and the technology to discover and describe all life on our planet. You will be pleased to know that your Swedish countrymen are right now at the forefront of doing this, starting with the fauna of Sweden. Already their investigations and discoveries have revealed a huge proportion of both new records for the Swedish fauna and species new to science. All of these discoveries are being documented and disseminated as widely as possible, and in a way that engages not only scientists, but people of all ages and from all social strata. I hope sincerely that very soon the rest of the world will follow the Swedish example which so clearly has its roots in your vast accomplishments. In conclusion, the very fact that we are gathering this week in Paris to honour you and your work, and to celebrate the continuation of your tradition, confirms that there is still hope that humankind will pursue the task of completing the enterprise that you began 250 years ago.

Sincerely,

Andrew Polaszek

Carl Linnaeus

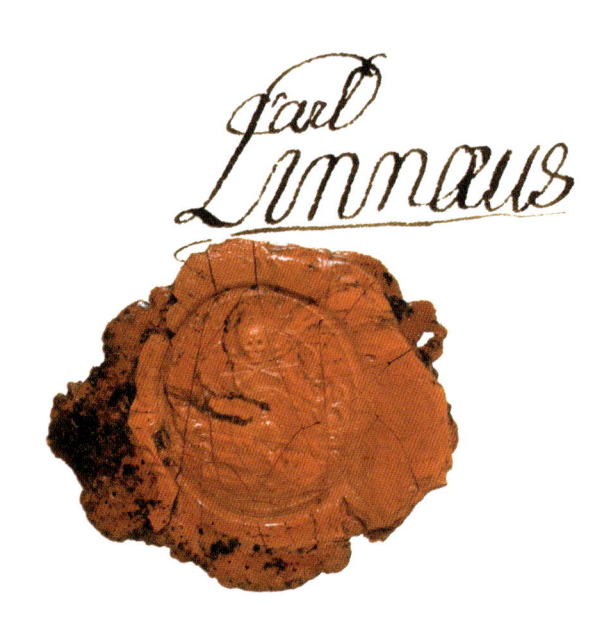

Illustrissimo Viro

Carolo Linnaeo

Maximas tibi gratias, Vir illustris, persolvo, quod Disser=
tationes Academicas mecum communicare volueris;
quasque mihi diligenter misit, Amicus meus fidelissimus
Dom. D. Garthshore: Donum mihi exoptatissimum!
Delecta praesertim Hedwigii, cum Famâ Methodi
Muscorum, nec non et commendationead meum usque
recessum, dudum volitabant.

A primis fere incunabulis, Plantarum amore captus,
Floram colui: Primum Ego, Raii nostri, nunquam non
venerandi, principiis imbutus; annum agens vicesimum,
antequam in meas manus pervenerunt Genera Plantarum
aliaque ejusdem Opera; inde statim lux nova; quid
haesitarem, quin viam monstraret Tuus Linnaeus.
erant exinde mihi Exemplaria Linnaeana, nocturnâ
versanda manu diurnâ: At vero vitam semper rure
degens, et Medicinam practicam exercens, extraque
fere commercium literarium positus, dolui iterum,
iterumque, quod nil in incrementum Rei herbariae
potui. Ne tamen Otium, mihi saltem, omnino infruc=
tuosum foret, anglicanas Plantas, praeprimis, non
penitus negligere decreveram; et propriâ manu
collectas, et amicis opitulantibus, ultra mille et
trecentas, siccas, in Herbario conservo.

Nuper etiam, ut nosti, Tui b. Patris illustrissimi
Marita, et scripta, ausus sum dicere, eo plane consilio,

Richard Pulteney (1730–1801) was an English botanist and physician. He writes to Carl Linnaeus filius [Linnaeus' son—Linnaeus himself had died in 1778] "Illustrious Carolo Linnaeo" detailing his botanical credentials and his honour and respect for the elder Linnaeus and his work. He thanks Linnaeus filius for a copy of the *Dissertationes*. He hopes his herbarium of English plants will be found to be of some use in the great work being undertaken.

ut apud nostrates, et discipulos maxime botanicos, Fama Viri b. amplius enitesceret, simul et Doctrina: Utinam Opus viro dignius! tale quod Otium, quod vires meæ potuerunt, quod Fontes, minime vero plenissimæ, recluderunt, quale, benevole, Tu ut accipias, et candide judicet orbis eruditus, plurimum optavi.

Si in quintuplo vel decuplo, habeas Murray Fundamenta Testaceologia, *des mihi unum precor, ni orbi publico cito reddituræ sint, in octavo Tomo Amœnitatum academicarum. Inter meas enim delicias, Conchylia etiam, aliquot annos, existimavi. Testarum species linnæanas sexcentas, ni fallor, et trecentas fere novas, ad normam Systematis dispositas, in musæolo servavi. Inter has, paucæ ab orbe novo conquisitæ sunt, in navigatione, quam sub sydere Galii, ad rem naturalem promovendam susceperunt Viri, jam per totum orbem celeberrimi.*

Tecum Lugeo, quod Horum unum, dum scribo fere, mors eripuit, Vestrum, intelligas, nostrumque eximium Solandrum!

Vale, et ut diu vivas, sospes, et incolumis precatur

Dabam
Blancofordæ,
apud Dorsettenses.
Augusti 27. 1782.

Tuus devotissimus
Ricardus Pulteney

He remarks on a book on shells *"Murray Fundamenta Testaceologia"*, and mentions his own museum of shells arranged to the Linnaean system. He jokes at the end that he must go, as he has been joyfully writing for an hour, like their mutual intelligent excellent friend Solander. The letter was sent from Blandford, Dorset on 27 August 1782, in a little over a year Carl Linnaeus the younger would die as well and the Linnaean collections would be bought by James Edward Smith and brought to London where they are now housed at the Linnean Society.

Dear Prof. Linné,

Please forgive my indulgence in writing to you directly. I have not yet had the honor of meeting you in person, and in light of the vast distance that separates us, I fear I may not enjoy such an honor anytime in the foreseeable future.

By way of introduction, I was trained in the art and science of your field by Dr. John E. Randall, a scholar of great accomplishment and repute who has discovered, documented, and assigned names to more species of fishes than anyone else in my century; including more species from coral reefs than anyone from *any* century! He was trained by Dr. William A. Gosline, a student of Prof. George Myers, himself a student of Dr. Charles H. Gilbert. Dr. Gilbert was mentored by David Starr Jordan, one of the most prolific contributors to our field, who in his early career was instructed by one of the greatest scientists of *his* century: Prof. Jean Louis Rodolphe Agassiz. Prof. Agassiz considered himself the intellectual heir of Georges Cuvier, the most famous naturalist in 19th century Europe, who is still regarded as having "possessed one of the finest minds in history". Baron Cuvier began his career at the Museum d'Histoire Naturelle in Paris through correspondence with Étienne Geoffroy Saint-Hilaire. You might know of Saint-Hilaire, who studied under your contemporary, Mathurin Jacques Brisson (author of such works as *Regnum animale in classes IX distributum sive Synopsis methodica*, and *Ornithologia sive Synopsis methodica sistens avium divisionem in ordines, sectiones, genera, species, ipsarumque varietates*). Saint-Hilaire was also a colleague of Jean-Baptiste Lamarck (a student of another of your contemporaries, Bernard de Jussieu), and took the position formerly held by Bernard Germain Etienne de la Ville-sur-Illon, Comte de Lacépède, who was inspired by the works of your colleague Georges-Louis Leclerc, Comte de Buffon.

However, perhaps our strongest connection to each other comes not from this pedigree, but through your dearest friend and colleague, Peter Artedi (Petrus Arctaedius). Let me here and now extend my condolences to you after Peter's most unfortunate and untimely death, and (as you know well) the great insights that were lost with him. In his loss, I am consoled (as should you be) by the fact that many of those insights were survived by you, not only through your determined efforts to secure and publish his manuscripts, but also in the way that your close friendship and frequent discussions with him helped to shape your own views on the classification of life, and hence the views of all practitioners of taxonomic classification who follow you through centuries! I also feel a certain kinship with Peter by our shared passion for all things Piscine. As with Peter, my fascination for sea creatures began in childhood, and this fascination has never left me (though I daresay I shall exercise utmost caution when strolling the streets of nocturnal Amsterdam, and will observe a similar degree of caution when engaging in contractual arrangements with the likes of Albertus Seba!). Today I continue my research on fishes and other denizens of the deep here in the 21st century, though I continue to read, learn from, and admire the accomplishments of many of our colleagues who span the two and a half centuries now separating us.

No doubt you have already been informed, through correspondence from my contemporaries, of the enduring legacy of your works. They have, I am sure, already explained how the basic principles you establish in *Editio Decima* of *Systema Naturae* are captured in a formal Code created by our esteemed colleagues in the twilight of the 19th century—a Code that has crossed more than a century to its present form (*Editio Quarta*), and remains as one of the longest-standing demonstrations of complete and widespread voluntary international cooperation within all of academic history, if not amongst all of human history!

I trust my contemporaries have also alerted you to the many amazing things we have learned through the centuries that separate us. Perhaps most significantly, one of the greatest scholars of the 19th century, by the name of Charles Robert Darwin (himself inspired by the works of Jean-Baptiste Lamarck), has shared his extraordinarily keen insights with humanity in the form of evolution by natural selection, thus providing a framework around which all scientific studies of life (including our field) are oriented. Darwin's ideas (along with those of his contemporaries; notably Alfred Russel Wallace), have revealed the utterly unbroken interconnectedness, through common descent, of all living things on Earth. Amazing though this revelation may seem to you, I assure you it is even more fantastic than you might first guess, as news of the true age of the Earth, and the span of time in which living things have inhabited its many environments, could not yet have reached you. Though you could not know it in your century, the hierarchical system of organizing Life you helped to establish provides what many scholars perceive as a near-optimal mechanism for representing the evolutionary affinities amongst biotic beings. (I will not concern you now with the imperfections of this perception, nor the reasons for the qualified 'near-optimal' descriptor alluded to previously; these are quibbles that we in the 21st century must contend with, and they do not in any way detract from the importance of your contributions across these centuries.)

There are some, even in this century, who find this notion of evolution scandalous to their world-view, and thus protest it (in spite of an overwhelming consilience of supporting evidence) as impious—in a

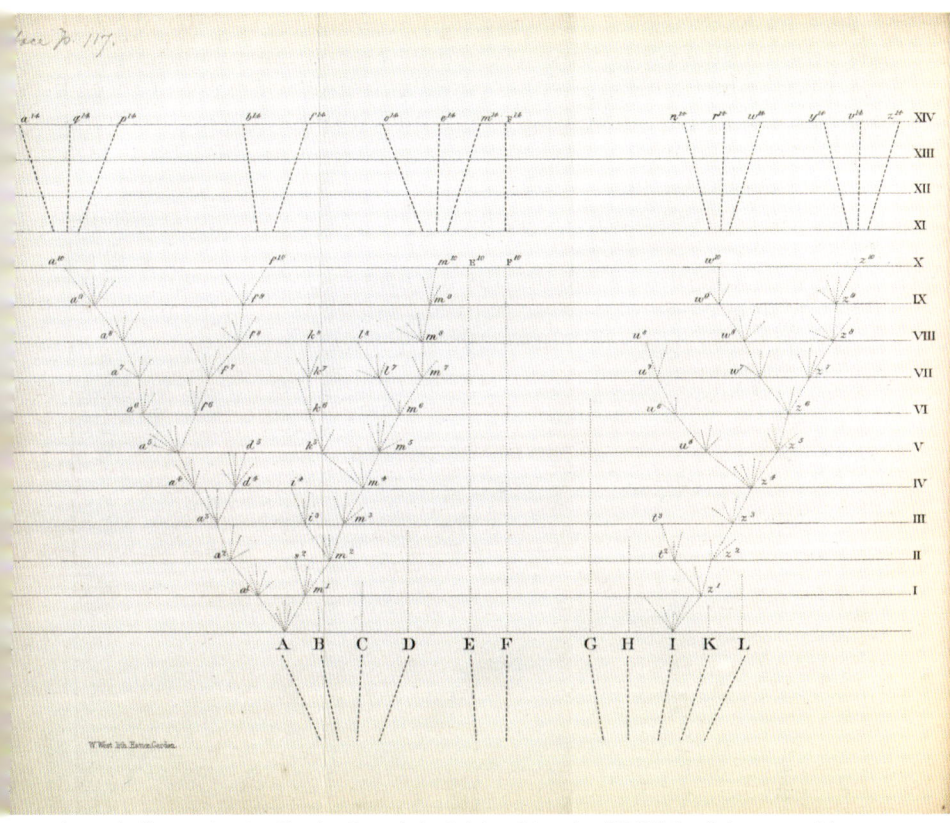

The only illustration in Charles Darwin's *Origin of Species* (1859) is of descent with modification through time

vein similar to accusations the Archbishop of Uppsala once directed at you. I do not wish to imply that the notion of evolution forces one to abandon faith in a higher spirit; for it most certainly does not (except, I suppose, to those whose faith is so fragile and tenuous on first principles). So do not be alarmed by this revelation; absorb its implications to your naturalistic world-view as it suits you (which I am certain you will do in a manner consistent with your amply demonstrated intellect).

In any case, I should explain my motivation for writing to you now. I have recently scrutinized every page of *Systema Naturae, Editio Decima*. These pages, and the important information they contain, have as much relevance in this century as they do in yours; perhaps even more so. (Few other published works of your century have demonstrated such longevity!) I have come to know these pages with such intimacy through an effort that my colleagues and I hope will continue to perpetuate your legacy for many more centuries yet to be explored. Alas, the original copies of your published works resist without perfect success the inevitable consequences of entropy (a concept of thermodynamics elucidated a century beyond your passing, which describes the trend towards disorder in all systems). We in the 21st century (particularly those among us practiced in taxonomic classification) understand and thoroughly appreciate the value of original scholarly works, and we seek to perpetuate them and protect them from the ravages of decrepitude. We do this not through some sort of modern alchemy of inks, nor advanced papyrus, nor any sort of incarnation of Johannes Gutenberg's invention. Rather, we achieve it through what we call 'digital imagery'.

This specimen of the Gudgeon (*Gobio gobio* L., Cyprinidae) from the Linnaean collections has suffered from both insect and physical damage

Rays of light reflected from the pages of surviving copies of your works are observed and recorded through the use of special instruments, and stored in a form that is far too small to be seen by human eyes, in a language consisting of only two characters (the numerals 1 and 0). We do not learn this language in our schools the way that one might learn scholarly Latin; instead we use machines to translate these recordings into a form that can be seen with tremendous clarity and vibrancy. So ubiquitous are these machines, that nary a student or scholar in this century would perform meaningful intellectual work without one. Perhaps most amazing of all, these machines are joined through thin copper wires and even thinner fibers of glass that allow the transmission of documents from machine to machine in no more time than is required for the blinking of one's eye. With barely a twitch of my left thumb, I can retrieve the pages of *Systema Naturae* from the other side of the World, and view them in such exquisite detail that I can see (and almost feel!) the texture of the sheets upon which it is printed. Through this miraculous innovation, we hope to allow generation after generation of scholars, through this and other centuries, ready access to your genius, for as long as humanity dominates our planet.

As I write this, the most direct manifestation of your continued legacy is found in the form of what we call *ZooBank*. Using our machines of instant global communication, we hope to create what is effectively the *Systema Naturae* of the 21st century. I am both humbled and honored to be playing an important role in the development of *ZooBank*, which explains why, as I commented upon earlier, I have paid such careful scrutiny to your *Systema Naturae, Editio Decima*. We have selected the date of 1 January, 1758—the date we establish for the publication of your important work—as the official start of modern zoological nomenclature. On this same date, a quarter of a millennium hence, I found great pleasure and satisfaction in flicking my left thumb upon my machine, thereby allowing global access to *ZooBank*. At that moment, each and every one of the 4,398 Species, 29 Subspecies, 315 Genera, 40 Subgenera, 34 Orders, and six Classes (as well as *Regnum* Animalia itself), as documented in your timeless work, were made available to the entire world.

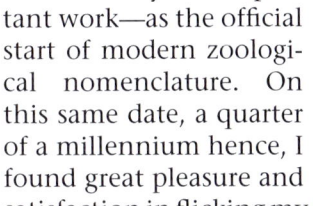

Chromis abyssus Pyle (Pomacentridae), the first species registered on ZooBank on 1 January 2008 (Illustration © Tamara L. Clark. www.tamaraclark.com)

You should be delighted to learn that, with the assistance of these same machines, we are continuing your legacy much more thoroughly than the mere perpetuation of your own published works. Indeed, we are continuing, with tremendous enthusiasm and as much vigor as our beneficiaries will allow, your Grand Endeavor to catalog all species. But work on this Catalog has not progressed at a pace commensurate with the grandness (and grandeur!) of the task. Alas, though the species you document within *Systema Naturae* represent all of Creation known to you, it is merely a symbolic start to an enormous effort yet to come. Through the two and a half centuries that separate us, our colleagues have expanded the listing of animal species some 400-fold over and above that which you include within your *Systema Naturae*. Yet, even our most modest estimates of Earth's total accounting of animal species exceed several thousand-fold above the diversity you document. Considering all forms of life, including the plants you know so well, in addition to an astounding array of organisms you could not have known existed as a consequence of their infinitesimal size (a realm of living diversity the immensity of which we are only beginning to appreciate here in the 21st century), the number of living species may well fill *ten thousand* volumes of *Systema Naturae*! The assembly of such an enormous Catalog is far, far too much work for any single person to accomplish (as I am certain you will appreciate), and so we require the cooperation and collaboration of naturalists the world over, and for many years to come.

Were it only a matter of patience and determination, I could rest comfortably knowing that two and a half millennia hence (following the current pace of progress) we might at last complete our grand task. But patience alone, in this case, does not afford sufficient virtue. It is here that I must, with great reluctance, inform you of another revelation in science first brought to light by the aforementioned Baron Cuvier. You are, perhaps, aware of strange rock-like bones discovered in China, Europe, and elsewhere; the tremendous size of which have led some to regard them as remains of Dragons. Surely you know of one such object, first commented upon in print by the Englishman Robert Plot in the century before yours, and assigned the name (in accordance with your methods) *Scrotum humanum* by your contemporary Richard Brookes, only five years beyond the publication of *Systema Naturae, Editio Decima*. (It is perhaps best that this name, so misaligned with the true nature, if not the superficial appearance, of the object it describes, has fallen into what our Code of your nomenclatural system refers to as *nomen oblitum*!) As it happens, Baron Cuvier very insightfully elucidates the correct Natural History of these rock-like bones as remnants of great beasts that once flourished thousands of centuries past, but which no longer walk the face of the Earth. We now know that the vast majority of all life-forms that have ever lived no longer share this planet with us.

Indeed, here in our century we are aware of at least five catastrophic episodes through Earth's history, in which tremendous numbers of species ceased to exist. Sadly, I must report that evidence is growing that here in the 21st century, we may well find ourselves at the dawn of the sixth such 'Mass Extinction' (as we refer to these episodes). With even greater sadness, I must confess that this horrible trend appears to be of our own making; the unintended but nevertheless tragic consequence of the increasing influence *Homo sapiens* has over our environment and the many other species we share it with. As the population of our species has now surpassed *ten-fold* that of your century, the diversity of life is being lost at a rate unprecedented in human history (perhaps unprecedented in *all* of Natural History!) You could no more have imagined this calamity, I am sure, than you could have foreseen our machines of global communication, and the many other wondrous inventions and luxuries we take for granted here. But I fear that news of this represents an even greater catastrophe than even *you* might imagine, for—thanks to the keen insights of the Augustinian priest Gregor Mendel, later investigations culminating with the triumphs of Rosalind Franklin, Francis Crick, and James Watson of the 20th century, and continuing in this century through the efforts of Craig Venter, Francis Collins, and many others—we are only just beginning here in this century to appreciate the diversity of Life for what it truly represents: no less than the greatest treasure this Earth has ever borne, or ever will bear in all of its future history!

The treasure of which I speak is manifest not in gold, nor jewels, nor any sort of monetary profit (though surely there is much profit to be made on the backs of living species!) Rather, this treasure takes a form that a scholar such as yourself can readily appreciate: *information*. The living things inhabiting this planet hold within them vast and precious secrets written in a form that we call the 'genome', and which we are only just beginning to understand. I will not trouble you with the details now (perhaps I can expound in future correspondence?), but suffice to say that each species is like a book, the contents of which have been refined and perfected over eons through the process elucidated by Darwin. Even in this advanced century, we are as mere school children running amid the halls of the Library of Alexandria, largely oblivious of both the value contained within the volumes that surround us, and the perilous fate that awaits it. Someday our species may come to understand how to fully unlock this treasure, but if we do not exercise sufficient care, this day may come after we have forever lost so many species—like books in a fire—along with the priceless secrets they contain.

But I still cling to optimism, challenging though things may seem to me now. I know there are others like you, like your dear friend Artedi, like the many brilliant scholars through the centuries that separate us, and (dare I suggest?) like myself—who are impassioned to complete your great adventure in documenting the many forms of life inhabiting this Earth. You are even now a major source of inspiration for those of us who follow in your very well-worn footsteps.

And so we continue.

I hope this letter finds you well, in good spirits, and resting in peace.

Aloha,

Richard L. Pyle, PhD
Department of Natural Sciences, Bishop Museum
Honolulu, Hawaii 96817
United States of America
21st century

Esteemed Carolus Linnaeus,

Let me start by saying that it's an honor to have the opportunity to write to you, one of the founders of systematic biology, my chosen field of study. I imagine that you're really busy with the preparations for the big anniversary of your *Systema Naturae* (10th edition), so please don't feel obliged to reply. As you certainly haven't heard of me, please allow me to introduce myself. I'm a research zoologist and curator of the collection of amphibians and reptiles (we now treat these taxa as mutually exclusive rather than nested) at the National Museum of Natural History of the Smithsonian Institution (which is old and respected in my country but did not yet exist in your time). I hope you won't hold it against me that I've chosen to study what you considered "pessima tetraque Animalia" (foul and loathsome animals).

I'd like to take this opportunity to clarify some things about a new approach to biological nomenclature with which I've been involved, because this approach is commonly but misleadingly characterized as an assault on conventions that you introduced. Contrary to this inaccurate characterization, the new approach is highly compatible with your own practices, particularly with regard to the function of taxonomic ranks. The approach I'm talking about is called *phylogenetic nomenclature* (I'll assume that other correspondents have filled you in on the concepts of evolution and phylogeny). It's based on the idea that taxon names can be defined with reference to common ancestry relationships, thus providing an objective method for applying names in the context of alternative phylogenetic hypotheses. Although this approach is commonly characterized as an alternative to the 'Linnaean system', nothing could be further from the truth. The reason for this misleading characterization is that the approach to which phylogenetic nomenclature is truly an alternative, which I prefer to call 'rank-based' rather than 'Linnaean', relies on the taxonomic ranks that you introduced (though dozens of additional ranks have been added to your original five). Nonetheless, calling the rank-based approach 'Linnaean' is highly misleading because it uses the ranks in a way that you never did and therefore causes names to be applied in ways that are at odds with the manner in which you and your immediate followers applied them. I should add that the rank-based approach was developed almost 100 years after the publication of the 10th edition of your *Systema Naturae*.

As an example of how the rank-based system works, consider the termites, which people have recently determined are nested phylogenetically within roaches. Because roaches (*Blattodea*) and termites (*Isoptera*) were previously considered mutually exclusive and ranked as orders, it's been proposed that termites be demoted in rank to a family of roaches (I'll italicize all scientific names, following the lead of both the rank-based botanical code and the draft phylogenetic code, commonly referred to as the *PhyloCode*). When this is done, the rank-based nomenclatural system requires that the name of the group of all termites be changed from *Isoptera* to *Termitidae* and thus also that the name *Termitidae* change its reference from a subgroup of termites to the group of all termites—even though the hypothesized composition of both of these groups has remained unchanged. And that's just the tip of the iceberg, because now all the former termite families have to be demoted in rank to subfamilies, and all of the former termite subfamilies have to be demoted in rank to tribes, etc., and all of these changes in rank necessitate changes in the names of the taxa that they designate (in particular, their rank-specific endings). With examples such as this in mind, it's hard to believe that the rank-based code used in zoology has a stated objective of promoting stability in the scientific names of animals!

The termite was classified by Linnaeus in his "Aptera", an insect group without wings

Of course, you and your immediate followers had the good sense to use ranks solely for taxonomic purposes (as opposed to nomenclatural ones), so that ranks had no effect on the application of names. As an example, consider the case of reptiles, which is more or less the reverse of the termite example. You had considered reptiles (*Reptilia*) to be a subgroup of amphibians (*Amphibia*) with the former ranked as an order and the latter as a class. In contrast, some of your followers (e.g., Macleay, 1821[1]; Blainville, 1822[2]; Latreille, 1825[3]) considered these taxa (albeit with some compositional changes) to be mutually exclusive and therefore elevated *Reptilia* to the rank of class. However, because ranks had no bearing on the application of names, no changes in the names of these taxa or the references of the names were required, and in fact, none occurred. (Some critics of phylogenetic nomenclature might argue that even today these names would not be affected by changes in ranks, but that's because the ranked-based approach hasn't been extended to ranks above superfamily, at least in zoology. On the other hand, there

de Queiroz 201

"Lacertilia" or Lizards (Table 79) in Ernst Haeckel's *Kunstformen der Natur* (1899)

have been proposals to extend the rank-based approach to higher ranks. Let's hope that such proposals never gain significant support!)

Phylogenetic nomenclature applies names similarly to the way in which you and your immediate followers did, because it uses methods that function independently of taxonomic ranks. By the way, I should point out that contrary to a common misconception, phylogenetic nomenclature does *not* attempt to eliminate your taxonomic ranks (and should not, therefore, be confused with proposals for rank-free taxonomy). It simply doesn't use ranks for applying names, thus effectively returning ranks to the strictly taxonomic role that they played when you introduced them. As a result, phylogenetic nomenclature eliminates nomenclatural instability that results solely from changes in ranks, though it permits (even requires) changes in taxon composition when hypotheses about phylogenetic relationships change. Consider the termite example, but this time let's suppose that the names *Blattodea* and *Isoptera* (and those of the various subgroups of *Isoptera* formerly ranked as families, subfamilies, etc.) had been defined using phylogenetic definitions based on previously hypothesized composition (e.g., *Blattodea* = the least inclusive clade containing the species previously considered roaches). In the context of the new phylogenetic hypothesis, these definitions would require only a single change in hypothesized composition and *no* name changes. The clade of roaches (*Blattodea*) would now be considered to include termites (*Isoptera*), but there would be no change in the name of the termite clade (*Isoptera*), nor would there be any changes in the names of its subgroups (i.e., those previously ranked as families, subfamilies, etc.).

These examples illustrate that phylogenetic nomenclature applies names to taxa similarly to the way that you (and your immediate followers) applied them. In both cases, taxon names have their primary associations with taxa (groups) rather than with ranks. The reason is that ranks function *solely* as *taxonomic* devices for representing hierarchical relationships (e.g., orders are included within classes) that have no influence on *nomenclature*, the application of taxon names (so that changing the rank of a taxon does not require changing its name). Of course, phylogenetic nomenclature also differs from your approach in using methods based on the principle of evolution (common descent). This difference seems highly appropriate given that most 21st-century systematic biologists consider the principle of evolution the unifying theory of our discipline.

I hope I've been able to convince you that phylogenetic nomenclature (and therefore the *PhyloCode*) is not at all an attempt to do away with your taxonomic innovations. Instead, phylogenetic nomenclature embraces your eminent wisdom in restricting the use of categorical ranks to the taxonomic function of representing hierarchical relationships. It frees ranks from their (later acquired) nomenclatural function by embracing the most important theoretical development since your time—the principle of evolution—which it uses to formulate precise and explicit statements concerning the references of taxon names without involving ranks. Thus, far from representing a challenge to your nomenclatural practices, phylogenetic nomenclature is more appropriately viewed as representing your own basic approach updated with evolutionary principles. Long live Linnaean wisdom!

I remain your humble disciple,

Kevin de Queiroz
Research Zoologist
Division of Amphibians and Reptiles
National Museum of Natural History
Smithsonian Institution
Washington, DC
United States of America

Notes:

[1] Blainville, H.-M. Ducrotay de. 1822. *De L'Organisation des Animaux, ou Principes D'Anatomie Comparée.* F. G. Levrault, Paris.
[2] Latreille, P. A. 1825. *Familles Naturelles du Règne Animal.* J.-B. Baillière, Paris.
[3] Macleay, W. S. 1821. *Horae Entomologicae*, Vol. I. Part II. S. Bagster, London.

Carl Linnaeus

Dear Professor Linnaeus,

It is a joy to be writing you and to have the opportunity to congratulate you on your very useful system for naming plants and animals. It is logical to retain the generic names that had become standard over the years, since those had become the normal way of referring to animals and plants from Classical times onward. As exploration proceeded, though, so many new kinds of organisms were found that it became cumbersome to invent new polynomial phrases for each of them. Depending on how many of them were described in a particular book, the individual 'names', or ways of referring to species, would change accordingly; and the only way of linking the descriptions in different books would be to repeat the complete references. Knowing that people would go on discovering distinct, new kinds of plants and animals for a very long time, one could see that the system would eventually become top-heavy and collapse under its own weight.

I know that you added the specific epithets only as a kind of short hand; after all, in your time, the number of books that described plants and animals were few, and one could refer to them all. Today we think with some longing at the time when you could spread out all of the useful references on a single desk-top, together with the pertinent specimens if you were fortunate enough to have them there, and form a compilation as you did in writing your encyclopedic works, *Species Plantarum* and *Systema Naturae*. As it turned out, subsequent classifiers found the binomial system that you had devised to be truly convenient, since the names for individual kinds of organisms could remain standard, changing according to principles that were developed over the following decades.

We also appreciate your beginning to form the units of higher classification that were to be elaborated over the years and form a way of collecting information about individual kinds of organisms. You could not have known how many kinds of organisms existed, since the exploration of the tropics had just begun during your lifetime and was soon to reveal biological riches that were unimagined in your time.

Nonetheless, the system of binomial names that you launched has proved durable. With the introduction of electronic databases the structure of classification and the means of adding new species have become more flexible, and the structure of language in referring to kinds of organisms suits the gigantic task that faces us. It doubtless will be formed more completely in the near future, so that we can get on effectively with the enormous task that you facilitated over 250 years ago.

In an age of mass extinction, one in which we expect more than ever from organisms in our attempts to reverse trends and achieve a sustainable world, those who classify plants and animals are playing an increasingly important role. Thank you for the tools you gave us long ago to assist in this task, tools that have proved durable and continuously useful in helping to spur and support our efforts to describe the natural world effectively.

Yours truly,

Peter H Raven

Peter Raven
Peter H. Raven, President
Missouri Botanical Garden
P.O. Box 299
St. Louis, MO 63166–0299, USA

Corvus corax L. (Corvidae), or the Northern Raven, from John Gould's *Birds of Europe* (1837)

Salve Imperator Plantarum!

Greetings from the distant future! As our temporal parts never overlapped, you never would have heard of me. But I have heard a lot about you. If I go down to the University, I pass by a statue of yours in beautiful oxidized copper, tipping my hat to you in admiration. Your spirit, wisdom, and wit have accompanied me through all my professional life, which I decided many years ago to devote to the study of systematics and evolution.

You may wonder what I mean by 'evolution' and how it relates to your *Systema Naturae*. In your time that term referred to the development of the embryo, but that is not how we use this term any more. In our time, 'evolution' refers to the transformation of species or the origin of new species from an ancestral one. A lot has happened in biology since your day, most importantly the general acceptance of a theory of evolution. It started with Charles Darwin, who published a first sketch of his theory in 1859. Successive versions were enriched with new theories about inheritance, the dynamics of populations, and the dynamics of species origination. Your famous system of plant and animal classification and nomenclature survived all those developments, in spite of the fact that you fell out of favor with evolutionary biologists. One of the architects of the Modern Synthesis of evolutionary theory, Ernst Mayr, wrote a book on *The Growth of Biological Thought,* which was published in 1982. In it, he accused you of having "destroyed the 'continuity of life' and replaced it with a hierarchy of discontinuities", in perfect accordance with your "essentialistic thinking" that insisted "on the constancy and fixity of

The bronze statue of Carolus Linnaeus in the Linnean Society of London is newly decorated with flowers every day

species." This, of course, was the result of interpreting you as an Aristotelian, and of interpreting Aristotle as a "watered-down Platonic typologist" (to use James Lennox's words). The reason is that you are said to have claimed that a species' identity is given by a species' *differentia(e) specifica(e)* within its *genus proximum*, which is the way Aristotle is said to have requested how a proper definition must be structured in his system of logic. These *differentiae specificae* were taken by the architects of modern evolutionary theory to be defining, i.e., essential properties of species, that is, properties of species that cannot change in any way at any time, and hence impart identity on species through space and time no matter how many varieties of a species might be recognized by systematists. In the wake of the Modern Synthesis, we had philosophers of science, such as David Hull, telling us that the Aristotelian, and consequently your own way of teaching systematics, has blocked progress in this field for 2,000 years. Darwin seems to have read your work in a somewhat different light, however, when he wrote in 1859: "The importance of an aggregate of characters, even when none are important, alone explains, I think, that saying of Linnaeus, that the characters do not give the genus, but the genus gives the characters; for this saying seems founded on an appreciation of many trifling points of resemblance, too slight to be defined."

I am afraid that worse may be in store for your legacy. There was another 'revolution' in systematics that was initiated by the German entomologist Willi Hennig. With two books he wrote in 1950 and 1966, Hennig showed systematists how to unequivocally group species into the historical entities we now call monophyletic taxa, or clades. Your distinction of species and genera as real entities of nature (*naturae opus*), families and orders as creatures of logic (works of *artis & naturae*) has served you badly in that context, since the disciples of Hennig consider clades as real entities of nature at all levels of inclusiveness. Cladists have forged an alliance with philosophers of science to declare your system defunct and hopelessly out of date. Pre-evolutionary in history and essentialistic in nature, your system is deemed to be incompatible with modern work in systematics and phylogeny reconstruction, which seeks to picture evolution 'as it really happened'. In the year 2004, an International Society of Phylogenetic Nomenclature was founded in Paris, with the goal to promote the PhyloCode as a replacement of, or at least a complement to, your own system. The irony of history is that while your system is rejected as being essentialist, the PhyloCode itself is essentialistic, albeit not in terms of an essentialism that originates from the nature of species, but from considerations in the philosophy of language (this is where the philosophers have helped out the cladists). It is no longer an intrinsic property of its constituent organisms that serves as the essence of a species, but a relational property instead: it is its unique evolutionary origin that is the essence of a clade, or species.

But here is the question I have for you! Would you, indeed, consider yourself an essentialist of the kind the Modern Synthesists and Phylocoders say you were? And if you did take clues from Aristotle on how to carve up nature into genera and species, would you agree that Aristotle was an essentialist of like kind? I know that the interpretation of Aristotle as such an essentialist is the classic one, but recently some philosophers have started to interpret Aristotle's biological writings differently. James G. Lennox, for example, in the second issue of the *Journal of the History of Biology* to appear in 1980, argues that for Aristotle, *differentiae specificae* needed not to be completely discontinuous, but that species could differ in degrees, i.e., in 'excess or defect', in 'the more and the less'. "Bird differs from bird by the more or by degree (for one is long-feathered, another is short-feathered", he reports Aristotle to have written. Brian Ellis, in his *The Philosophy of Nature* (2002), points out that for Aristotle, an animal born as a tetrapod does not cease to be a tetrapod when it loses a leg in an accident, or misses one as a birth defect, but continues to be a tetrapod because this is its nature, and its nature is determined by its development. Ellis interprets Aristotle for having believed in intrinsic causal powers of development, which explain why 'like begets like', i.e., why a fly does not beget an elephant.

And as for yourself, did you not argue that we must learn with great diligence and attention to read the 'signs' that give the genera and by which we must distinguish species? These characters are not mysteriously 'given' to us to be used in the definition of species; they must be discovered in order to arrive at a correct description of species. If it is true, as you said, that there are cases in nature that require great experience and the knowledge of a genius to properly distinguish a 'variety' from a 'species', then you must have implied a difference in degrees, rather than in kind. The same implication would seem to obtain from your remarks on *Homo troglodytes*, as you considered it uncertain whether he shares closer affinity with the pygmy or with the orang-outang. You did allude to the Great Chain of Being, but rejected its depiction as a ladder, for you acknowledged that although some species share a striking family resemblance, others appear separated by a discrete gap. You painted the picture of a web of nature to capture the family resemblances appreciated by the systematist, but the web would be torn in places where gaps seemed to prevail. Another metaphor you proposed was that of a geographical map: some families reside in the same country, others in neighboring

The "scale of being" depicted as a ladder by Charles Bonnet in *Traité d'insectologie* (1745)

countries, others on islands. Where gaps might suggest discontinuity, family resemblance suggests difference by degree. But if the web of nature seems torn, if neighboring countries or land-bridges appear to be missing, this still leaves open the question of whether such discontinuity marks essential difference, or human ignorance. Students—not teachers—of nature, as you said we all are, have to proceed from the special to the general, hoping that such method will help us bridge the puzzling gaps by the discovery of missing links.

I am sure most people who know your work also know that in your later career you argued that plant species could originate through hybridization. Rolf Löther, in his book on *Mastering Diversity* (1972), argues that you considered genera and species to be the work of the Creator (*Creator telluris omnipotens*), and of its executive arm, i.e., Nature (*Exsecutrix ejus*). If plant species originate through hybridization, they become daughters of time (*filiae temporis*), which simply means that the executive branch was assigned more tasks to complete according to the Plan of Creation. But if plant species are children of time, how can they be marked out by eternal unchanging, essential properties that impart identity on species of plants and all their varieties as they result from their natural growth, or artificial cultivation, under different environmental conditions? That would appear to be paradoxical to me. And so, I assume, it would have been for you.

My dear Linnaeus, it is late now, and I must stop. It was wonderful spending an evening thinking of you and your work. And if, as seems to be the case, the Modern Synthesis of Evolutionary Theory mis-represented your thoughts, rest assured that historians of science are working to correct that picture. One of them is Mary P. ("Polly") Winsor, who in her 2006 contribution to the *Annals of the Missouri Botanical Garden* argued that the Modern Synthesists and their associated philosophers were busy creating an 'Essentialism Story', an "urban myth" as she calls it, just so they could have a straw man to knock down. Your legacy still matters to many of us, and we strive to get it right. So rest in peace!

Cordially yours,

Olivier Rieppel
Department of Geology
The Field Museum
Chicago, IL 60605–2496

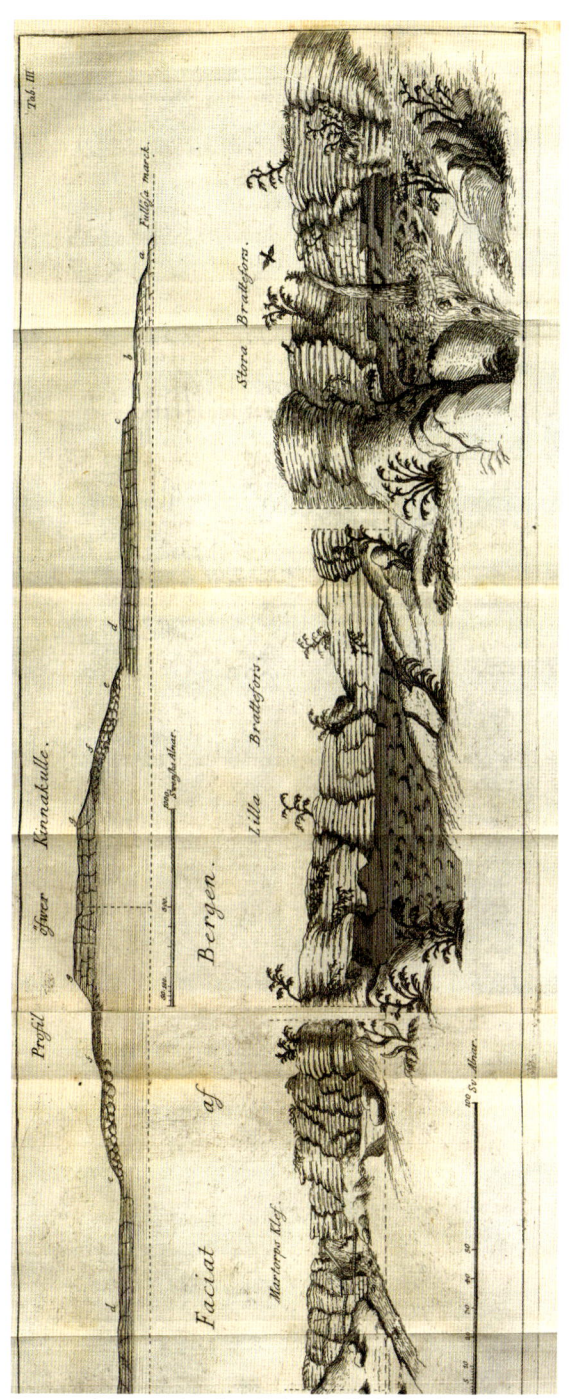

A cross-sectional map of the peninsula of Kinnekulle in Götland (south western Sweden) drawn by Linnaeus during one of his few field journeys

Carl Linnaeus

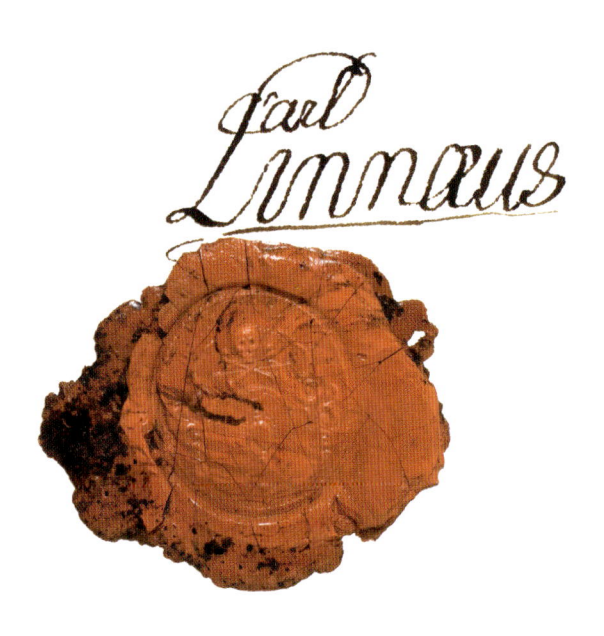

They call him 'L.'. It means 'the'.

Dear Carl,

It must have been the best part of 55 years ago. Early 1953, I guess, and I'd have been coming up to four years old. I remember the frost. Crisp, sharp frost, the grass iced white, and the crunch of my father's Wellington boots as he carried me on his shoulders down the path and into the dip at the end of the garden. The dip had once been a gravel-pit but it had been abandoned long ago. The sides had collapsed to gentle slopes of hazel scrub and bramble. The centre provided a grassy arena some twenty yards across. I remember watching my legs, snug in padded dungarees, and the little black boots on my feet (just like Daddy's big ones) bounce with each of his steps. He was singing as he went—it was always the same song, and I have never known whether he just made it up or whether it was a fragment of some long-lost country hit from the 1930s: "Carry me over the mountain ridge, down the trail to the old pine bridge"—and then there were a couple of mumbly bits which were his admission that that was all he could remember. Juno padded behind, her breath steaming in the cold air, black Labrador on white frost, tail wagging. This was our time together, our morning ritual. Emptying the trap.

"There won't be much this morning," he said. "Too cold." I was too busy to reply. I was making puffs of steam, just like the shunting engines that herded the grain trucks in the railway sidings up the road by the animal-feed mill. Where I wasn't supposed to go. But Juno and I had a guilty secret. The previous autumn we had burrowed under the fence, and one thing led to another. Now we would go train-spotting on a regular basis—under the fence and the hedge, up the road almost to the bridge, up the bank, by-passing the station entrance, and emerge at the end of the platform and watch the action. The station-master was a tolerant man, and this was before the days of the Health and Safety Act. One day we had actually got to ride in the cab of one of the shunters. And on our next excursion I had taken my little red seaside spade with the wooden handle and helped the fireman. And Mum still hadn't twigged to what was going on. Despite the coal-dust.

Dad set me down on the ground and opened the haversack with the glass-topped boxes in case there were any 'finds' to take indoors for the collection. He carefully lifted the lid of the moth trap, checking the underside for moths before setting it down on the grass. Then out came the rough papier-mâché egg-trays. "One for me and one for you. Be careful." He was right—there wasn't much. I turned the egg-tray over, being careful not to drop it nor to squash anything. There were a few spring things, and I knew the names of the common ones, *gothica, incerta, stabilis*. There was nothing new. We swapped trays; Dad inspected mine and I inspected his. Then we marched in unison to the side of the gravel-pit and shook the moths into the bushes. Back to the trap. "One for me and one for you. Be careful." Two miserable noctuids. I turned the tray over. And my heart stopped. It was simply huge, great velvety purple-brown curtains of wings with toffee and chocolate and cream, and an enormous body with a big yellow face on it. "Dad?" "Wait a moment, my boy." "Dad!" "Hang on." "*Daddy*!!!" "Good Lord! What do we have here?" "Whatisitwhat'sitcalleddoyouwantit?"

Dad re-packed the trays in the trap, put the lid on and led the way home. I followed, carefully carrying the prize in the largest glass-topped box that we had, using both hands. Back into the warmth of the cottage, across the red-tiled undulating kitchen floor, along the passageway and into the study. My father carefully inspected the abdomen. "Male," he said. "Pity—we could have had eggs if it was a female," as he held the box open above the cyanide bottle and gave it a sharp tap. The moth dropped and he clamped the lid firmly on.

He opened a glass-fronted bookcase, one of the three that grace my study today, pulled out his copy of Richard South's *Moths of the British Isles* and turned a few pages. "Look," he said, "that's what it'll look like when we've set it." "What is it?" I asked. "*Atropos. Acherontia atropos*. Some people call it the Death's Head Hawk Moth."

Acherontia atropos L. (Sphingidae), or the Death's Head Hawkmoth

"Why has it got two names?"

We had always used what my father called 'Latin' names for moths, but in mixing with other naturalists (which Dad did, and I was his shadow) there had been inevitable exposure to vernacular names. I had never questioned this before, but *atropos* brought it to the fore. I knew that Latin names were made up of two bits and, within the context of British moths, the last name was almost always sufficient for everyone to know what you were talking about. And there is nothing precocious about using Latin names when you are that age—especially when everyone else uses them. I was even aware then that there were advantages of brevity—the vernacular names of some British moths are not only arcane, but also tongue-twisters, and '*varia*' (it's *porphyrea* nowadays) is infinitely easier than

'True Lover's Knot'. As for 'Blair's Shoulder-Knot', we're not going there, Carl.

Dad laid it all out as simply as he could, and explained that all over the world, because they all spoke different languages, people had different names for the same animal or plant. That came as a bit of a revelation until I thought about it and realised that when my parents said "Pas devant les enfants" and launched into incomprehensible gibberish they were actually *speaking a Foreign Language*. And they were probably doing it so that *I couldn't understand them*. I resolved to Learn a Foreign Language when *I* went to school (as it transpired, *that* wasn't as simple as it sounded, Carl). Until about two hundred years ago there were no names that everybody could use such that, when they did so, they could be understood in any language. And then a man called Linnaeus invented all these names for all the plants and animals that he knew, and since his time more people have discovered more plants and animals and they have given *them* Latin names—now there are *thousands and thousands* of names. And there are names for species (you know what those are—like *atropos*) and names for genera (you know what those are—like *Orthosia* and *Acherontia*), and names for families (like Noctuidae) and names for other groups, but that's quite enough and it's breakfast time. And so I sat at the breakfast table and practised saying *"Acherontia atropos"* with a mouth full of porridge.

My father nearly died from pneumonia in the winter of 1953–1954. He put himself on the transfer list for service in what were then the colonial territories and drew Singapore, which was hardly the short straw. So I found myself learning a whole new lexicon—*memnon*, *helena*, *agamemnon*, *chrysippus*, *diocletianus* and the near-mythical *brookiana* among the butterflies, and *gangis*, *atlas*, *maenas*, *nessus* and *celerio* among the moths. And I lived among people who *all* spoke a Foreign Language. We had the Indo-Australian moth volumes of Dr Adalbert Seitz's *Macrolepidoptera of the World* so I was able to identify a small but encouraging proportion of the moths that we caught. And I eventually learned an important lesson—that not all text-books are comprehensive or user-friendly, and not all species have been described and named, and there are some groups where identification to species is simply not possible. One just has to accept a genus as

Attacus atlas L. (Saturniidae), or the Atlas Moth of south east Asia

the next best thing, so I learned to write things like "Unidentified *Cleora* species" in my best joined-up writing.

As the years went by I learned more about scientific names and how they were used—and misused. I learned about synonyms, those 'oops' happenings in taxonomy where someone who should know better describes and names a species that has already been named. And like everyone else I was repeatedly confused by how species had, over the years, been tossed lightly from genus to genus with changing opinion and changing fashion, leaving a sometimes very muddy bibliographic trail. I learned that not just the genus and the species name were important, but sometimes one had to attach the name of the author and the date that he/she described the species to be sure of what one was really talking about. And I learned to be wary of subsequent changes made to the original name that the author had proposed— changes of termination of species-group names to adapt them to the 'gender' of the genus, and changes where a subsequent author simply thought the name wasn't spelled correctly or didn't sound quite right. Or simply couldn't be bothered to get it right. My learning was in no way structured—it just became intuitive after using the tools—the plethora of books and monographs and papers that have appeared since your day, Carl.

Once I had reached secondary school, my formal teaching in biology had contained very little about the 'how' or the 'why' of identifying and naming animals and plants. This surprised me a little, for it seemed to me to be a no-brainer that the correct identification and naming of an organism was a prerequisite to saying anything meaningful about it. My teacher was, mercifully, a committed botanist in her spare time and understood the inevitable conflict between a geek who had to have an accurate label for everything, and the demands of the Nuffield syllabus that required one to have a profound understanding of the molecular basis of photosynthesis.

I was 16 when dear old Dad, older and greyer and still just as devoid of musical ability, had pulled off the ultimate overseas service coup and landed himself the job of setting up a social security system in Fiji. And I was able to join him and my mother for two of the three annual school or, later, university vacations. We set up a moth trap in the garden and eventually built another six monsters from 44-gallon oil drums and positioned these strategically around the main island in every habitat from cattle rangeland to cloud forest. It was moth-hunter heaven—a practically unworked isolated island fauna full of things we had never seen before. And neither, we were to learn, had anyone else seen many of them either. And Dr Adalbert Seitz's tomes were precious little use. On that first holiday I collected Macrolepidoptera, sorted them to what looked like species, and made up a synoptic collection to take back to London. My father had looked at me over his spectacles and said, ponderously, "You had better go and see young Bradley at the British Museum and ask him what to do." So some weeks later I found myself within the hallowed portals of what is now the Natural History Museum in London. Then it was the British Museum (Natural History), but still known universally as the 'British Museum' or just 'BM', a homonym of its parent organisation in Bloomsbury.

I remember forbidding warders, gigantic Chubb keys that opened tall mahogany doors, and a smell that was to become an evocation for life—naphthalene, camphor, thymol, paradichlorobenzene, oil of creosote, a cocktail inhalant of traditional museology. Offices with neat nameplates on their doors ran alongside a cavernous, echoing and, at first glance, almost entirely uninhabited collections hall. At the far side of the rows and rows of cabinets were the 'bays' where the officeless lived in little personalised enclosures of bookcases and filing cabinets and open-fronted racks for drawers of insects.

John Bradley had accompanied my father on two expeditions to southwest Ireland in the very early 1950s. He had persuaded the Museum to stump up to send him and his wife on honeymoon to the Solomon Islands in 1953 to collect Microlepidoptera; he knew his Pacific and, just as importantly, knew who was doing what with Pacific Lepidoptera that could be mutually advantageous. He was no longer young, at least to me. But, as I soon learned, he was one of the least forbidding of the BM lepidopterists, and he was mightily encouraging. He introduced me to some of the staff of the Lepidoptera section—Alan Hayes, Steve Fletcher, Allan Watson and Michael Shaffer among many others. As John was concerned mainly with Microlepidoptera, he tried very hard to persuade me that these would repay study far more than the Macrolepidoptera. However, I was still a long way down on the learning curve and reluctant to bite off more than I could chew. Eventually, I ended up beneath Alan Hayes's wing. Alan was curator of the Noctuidae and he was a gentle and knowledgeable mentor. And then there was Timothy Tams. Tams was round-faced and rounded-backed, a small, bald, smiling, old man who was a walking encyclopaedia of the Lepidoptera. I had first met Tams when I was a very small child, and for Tams no time had passed. He was a legend in his own lifetime, not a prolific author, simply an oracle. He had written the moth volume of *Insects of Samoa* 60 years before I began work in Fiji, and his knowledge was invaluable. He had known George Hampson, Hampson had known Henry Stainton, Stainton had known James Stephens, Stephens had known William Turton, Turton had known Johann Christian Fabricius and Fabricius knew you. Tams was my link to you, Carl—seven degrees of separation.

Within a couple of years I was a regular visitor to the BM's Lepidoptera Section and began to soak up the arcane knowledge that took me to new levels of understanding of how systematics and nomenclature worked. My mentors introduced me eventually to that set of tablets that now govern how we use your names, Carl. I refer of course to the *International Code of Zoological Nomenclature*, a dizzying document that seemed (and often still does) to be a set of rules that dealt mostly with exceptions to those rules, and how to deal with the exceptions. But for the most part the basic provisions and precepts have protected 'your' system of names and ensured that 250 years on it still works. And you would perhaps be surprised at how many names there are now, and the methods we have to keep track of them. No doubt your other correspondents will tell you more. We have problems with remarkably few of the fundamentals, but the requirement for gender agreement (species-group name must agree in gender with generic name) sits most uneasily with the use of computers for cataloguing species. I'd take time off to explain computers, Carl, but we'd have to start with salt-and-metal sandwiches and frogs' legs, then work our way up—it might take a bit of time. And gender agreement sits uneasily too with the *Homo sapiens* who use and work with the names—they are no longer versed in Latin and Greek. In fact, much taxonomy nowadays is done using Russian, Japanese, Chinese or Spanish. Usage will eventually change, I am sure, to the original author's original spelling. Indeed, there has been a maverick movement among workers in some groups to pre-empt that change, and I am one of them.

Back in the 1960s when I really started getting into taxonomy it was decidedly typological. There were a few crazies out there coding characters and morphometric data as numbers and clustering things by degrees of similarity using computers that cost millions of dollars and were powered using coal, steam and valves. But numerical taxonomy didn't last long, although it didn't half sharpen up a few taxonomists' numeracy levels and improve immeasurably our use of statistics and such mathematical tools as multivariate analysis. But my first paper describing new species was done much in the way it was done in your day—use a reference collection, use books, compare specimens minutely and describe similarities and differences. And we talked about "closest relatives", "nearest allies", and the like. Intuition took us a remarkably long way. And we used names formed and manipulated in the Linnaean way.

You would be blown away by today's reference collections, Carl. These are still the basis of all that we do. Back in your time the size of collections was measured in hundreds and occasionally thousands of specimens. You and your contemporaries thought in terms of a world inhabited by a few thousand species. Now we think of there being more than three million species and the sizes of our major reference collections are measured in millions of specimens. And the technology of collecting and preserving has changed beyond all recognition. We can rely upon fungicides and insecticides to protect specimens collected in the wet tropics until we can get the specimens into air-conditioned and dehumidified conditions. And with flying-machines that carry people and cargo we can get specimens from the East Indies to London in less than 48 hours. Light traps, Malaise traps and interception traps have revolutionised the quantities we can collect. But the format of collections has changed little. We still pin most insect specimens, and put data labels on the pin, much as you did. Once the specimen is identified and named it goes in a drawer above a name label giving us genus, species, author and date. The first time I saw one of my new species so immortalised it was the biggest ego trip imaginable. And that, of course, is the problem with much of the taxonomy that has been done since your day. It has been an ego-trip for the describer rather than a discovery trip for the reader of the publication. Roger Crosskey coined the term 'mihi-itch' to describe the syndrome of which the symptoms are self-glorification for the describer of a new species and frustration for those who have to interpret and use the work afterwards. And creating taxonomic publications for the benefit of the user-community rather than the author involves a moment of revelation, a Damascus moment, that for some taxonomists never comes. So to those who have tried to use my pompous and mihi-itchy early publications, all I can say is "Sorry! I didn't know any better at the time." And to those mentors such as Klaus Sattler who were rough enough to throw me in the ditch on the road to Damascus and give me a sharp slap on the head, I say "Thank you. NOW I know what you were talking about."

But back in the late 1960s I was hooked. I resolved I would become a taxonomist. And I worked at it. I collected assiduously and learned my way around the collections in South Kensington, I wrote more about the Pacific moth fauna and eventually persuaded my university to support a taxonomic PhD. All the time, behind me, was that great corpus of taxonomic knowledge embodied in the publications of the last 250 years and in the great reference collections. And the key to navigating this great sea of knowledge was binomial nomenclature. Your system, Carl. Genus, species and author. Two of those authors stood out for almost as long as I can remember. You and Fabricius. Why? Because you were so famous, so well-known, that you didn't get your name spelt out in full. You were 'L.' and he was 'F.'. Everyone else was given full name, although within the microcosm of Lepidoptera we shortened up a few of the more unwieldy for day-to-day purposes—'H.-S.' for Gottlieb

August Wilhelm Herrich-Schäffer and 'Wals.' for Thomas de Grey, 6th Baron Walsingham, for example. But there was dangerous inconsistency in these abbreviations; I believe you and Fabricius got permission to be abbreviated in the Code, but no-one else was so honoured.

In the fullness of time I became a member of the Natural History Museum's staff—a fully fledged but still very green professional taxonomist. And our science was progressing—phylogenetic systematics was being developed. The promise of computers was being realised. But still your system held it all together. It was remarkably robust.

In the early 1980s I was privileged to work with the late Ebbe Nielsen on your collection of Microlepidoptera, Carl, to go back to your original specimens and check, for example, that what we call *Cydia pomonella* (L.) is really the species that you described and named back in 1758. Does it surprise you, I wonder, to learn that almost all of your specimens still exist, and they are in remarkably good condition, considering the years that have passed? It would suprise you as much, I imagine, to learn that your collections and library are in an air-conditioned strongroom in London, treated as very precious objects. How and why they got to London and the basement of the Linnean Society is a long story for another day. By and large, what we found matched the usage of your names today. Was our educated guess about the meaning of what we called the 'n'-labels on your specimens correct? These were the little labels with a lowercase 'n' and a numeral, like 'n116'. Were they

Cydia pomonella (L.) (Tortricidae), or the Codling Moth from the Linnaean collections

The labels from the specimen of *Cydia pomonella* in the Linnaean collections

really Lusitanian specimens from Domingos Vandelli in Lisbon, as we suggested? I hope so—we haven't had that theory blown out of the water yet, and it's been 25 years.

Along the way we found your friend Carl Clerck's long-lost collection of Microlepidoptera, unrecognised for over 200 years in the backrooms of the university in Uppsala, mixed in with Leonard Gyllenhal's collection. Yes, that Carl Clerck, the one who tried to illustrate the Lepidoptera in *Systema Naturae* and exhausted himself and his resources in just a few years and, sadly, died in 1765.

I was going to take you to task about *Phalaena* (*Tinea*) *vestianella*. Yes, your *vestianella*. The clothes-moth, I know. What's wrong with it? Well, it turns out that the name just didn't get used very much after *Systema Naturae* and practically everybody used your other name, *pellionella* (yes, I know, the other clothes-moth, the case-making species) until Arvid David Hummel introduced the name *Tineola bisselliella* in 1860 for the species that just makes a mess of silk and doesn't make a case. Yes, I know you got there first with *vestianella*, but hold on. Your original specimens of *vestianella* got lost or destroyed, yes? And you thought you'd better replace them so you pulled three specimens of a little scuttling pale moth off the windowsill, gave them a whiff of laurel and pinned them over the *vestianella* label, yes? I thought so. Well, Carl, you remember a week or two before that you'd brought in some *Atriplex patula*? Stop blushing, we all know what the seeds are used for: "A pound of these bruised, and put into three quarts of spirit, of moderate strength, after standing six weeks, afford a light and not unpleasant tincture; a tablespoonful of which, taken in a cup of water-gruel, has the same effect as a dose of ipecacuanha, only that its operation is milder and does not bind the bowels afterwards...". Got it in one,

Atriplex patula L. (Chenopodiaceae) – the Common Orache or Spear Saltbrush (LINN 1221.19)

Carl—your specimens are a coleophorid that emerged from the *Atriplex*, not the webbing clothes moth. So rather than upset an established name for a major pest we now have *Coleophora vestianella* (L.). I hope you don't mind, but we thought it was in the interests of stability to let that sleeping dog lie.

Another 25 years on from those days poring over those specimens of yours and I am starting to wind down. From drawing with Indian ink and typing our manuscripts on typewriters we have moved in those 25 years to digital imaging and word-processors, even automated dictation. We've got data of every kind on all sorts of organisms all over the internet, DNA and RNA sequences, images of hundreds of thousands of plants and animals, millions of words, kilometres of library shelving, databases that attempt to list every species on earth. We can reconstruct phylogenies that appear to reflect evolutionary patterns (another thing that takes a bit of explaining—we start with variation in domestic pigeons and work up). And amidst all this electronic wizardry we still do the things you did—we look at whole organisms, we take them apart to see how they compare, we describe what we see in terms of difference and similarity, we place them in a system of classification, a *Systema Naturae* that would be familiar to you. Twenty years ago, there were those who claimed that 'Linnaean nomenclature' would not or could not cope with the changes. Guess what?

So that is your legacy—an extraordinarily simple concept that seems to be able to cope with ever deepening levels of complexity. Yes, amidst all this electronic wizardry we still find species by the names that you designed for them, Carl. And, somehow, I think that we always shall.

They call him 'L.' It means 'the'.

[With apologies to *Once Upon a Time in Mexico*, 2003.]

Carl, you are The Man.

Kindest regards,

Gaden Robinson

A Paris le 21 7bre 1771.

Printed in Linn. Corr.
v. 2. 552.

Recevez avec bonté, Monsieur, l'hommage d'un disciple très ignare mais très zélé, de vos disciples, qui doit en grande partie à la méditation de vos écrits la tranquillité dont il jouit, au milieu d'une persécution d'autant plus cruelle qu'elle est plus cachée, et qu'il couvre du masque de la bienveillance et de l'amitié la plus terrible haine que l'enfer excita jamais. Seul avec la nature et vous, je passe dans mes promenades champêtres des heures délicieuses, et je tire un profit plus réel de votre philosophia botanica que de tous les livres de morales. J'apprends avec joye que je ne vous puis pas être à fait inconnu et que vous voulez bien même me destiner quelques unes de vos productions. Soyez persuadé, Monsieur, qu'elles feront ma lecture chérie et que ce plaisir deviendra plus vif encore par celui de les tenir

Jean Jacques Rousseau (1712–1778), the French philosopher, was greatly influenced by Linnaeus' book *Philosophia Botanica*, published in 1751. In this book Linnaeus laid out his principles for identification and classification "so that the terms and principles might be explained in one work, which Linnaeus considered as a matter of importance, not only to the learned world but also to his pupils; wherefore the work was concluded".

Rousseau, writing from Paris on the 21st of September 1771, asks Linnaeus to receive with joy the homage of a disciple of his disciples, who owes to his [L's] writings the "tranquillity" in which he works. He tells Linnaeus that his promenades in the countryside, alone with Nature and Linnaeus, are delicious hours, and that he has taken more profit from the *Philosophia Botanica* than from all books about morals [philosophy]. He tells Linnaeus how, in his "aged infancy" he is making a collection of fruits and seeds . . . he ends his letter with great praise—"Adieu Monsieur, continue to open and interpret for men the book of nature . . . I read your works, I study them, meditate upon them, I honour and love you with all my heart".

Dear Professor Linnaeus,

I have finally worked up the nerve to write to you after having studied your works for many years. Although it has been a very long time since you ceased publication, I would like you to know that your fellow naturalists and scientists still think very fondly of you. We look in awe at your astounding output and the influence your publications continue to have today. You had already published the *Systema Naturae* when you were 28 years old. When I think of what I was doing when I was 28 years old . . . well, perhaps it's better not to think of that.

The world is different in many ways from when you were active. Women are encouraged to study and to have careers in the sciences now, although there are still far fewer of them than there are male scientists. After reading your books and articles, as well as many books and articles about you, I wish I had the opportunity to meet you in person. Would you be interested in discussing botany with a woman from Canada? Would it have made you uncomfortable to know that I live alone at 31 and have no children? As for me, I would have liked to attend your lectures. I heard that they were not just informative, but also very entertaining. I would have liked to have seen your eyes sparkle as you made off-colour comments, too. (Don't worry—these days, even women from good families are not as sheltered as they were in 18th-century Sweden.)

You would probably be impressed but not surprised at how the system of nomenclature that you made popular is used today. It is certainly helpful to scientists in their working lives and to recreational naturalists in the field. I have also found it useful in my personal life. Last year I was invited to speak about you at a conference in Mexico. Latin has long fallen out of favour as the language of scientific discourse, replaced by English. I don't speak Spanish, and my hosts did not know the English names for many of the food plants we were eating, but we could talk perfectly coherently about them using their Latin binomials.

Speaking of tropical fruit, the banana you had the good fortune to taste from the *Musa paradisiaca* plant you coaxed into flowering in Clifford's garden is no longer considered exotic, even in northern locales. Although nobody has succeeded in acclimatizing bananas to temperate regions, advances in transportation and food preservation allow fresh bananas to be shipped all over the world without spoiling—and they are often less expensive to purchase than locally-grown apples. Canned pineapple (*Ananas comosus*, though you called it *Bromelia ananas*) can also be bought just about anywhere, even in remote Swedish towns. I know you have always been in favour of each country being self-sufficient in food, energy, and other resources. I suspect you would be appalled at the idea of a global economy based instead on the international trade of commodities manufactured where costs are the lowest. I am willing to bet that the thought of stores in Sweden selling bananas from Costa Rica or Turkey would horrify you. However, in the last few years, many countries have begun to rediscover the importance of self-sufficiency, particularly as the global prices of imported food has been rising due to the increasing cost of fuel. Some regions that focused on growing cash crops for export at the expense of producing staple foods now find themselves pressed to feed their citizens. Politicians are now at least talking about changing government policies to encourage farmers to grow crops to be sold locally, and it has now become trendy in some of the richer countries to choose locally produced food over imports. All this means that more research is being done to find varieties of food crops, livestock, and fodder that can grow in particular climates. I think you would be pleased with this development. Nevertheless, I think you would also like the opportunity to buy inexpensive chocolate whenever you want—you are, after all, the person who called the cocoa plant *Theobroma*, or 'food of the gods'—and to taste persimmons (*Diospyros kaki*—named by your son) or sliced santol (*Sandoricum koetjape*) with chilli. International trade does have its benefits, after all.

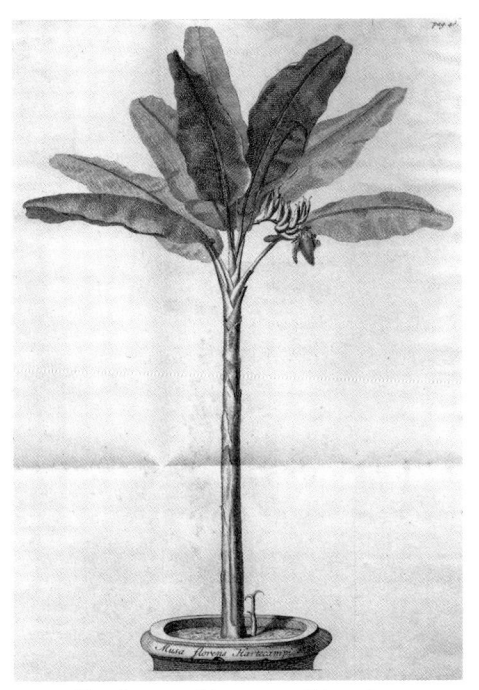

The Banana *Musa paradisiaca* L. as illustrated in Linnaeus' *Musa Cliffortiana* (1736)

As you can see from the names of fruit above, even though we still use Latin binomials for generic and trivial names, many of the names of organisms we use now are derived from languages other than Latin or Greek. For instance, 'kaki' is the Japanese word for persimmon. Quite simply, there are not enough words in Latin or Greek, or enough names of naturalists, patrons of science, or mythological entities, to supply distinctive monikers for the millions—yes, millions—of species we now know to live on this earth, not to mention fossil species. (You may also be shocked to hear that the language with the largest number of speakers these days is Chinese, and, believe me, Chinese speakers find Latin names quite unpronounceable!) Can you imagine that the Coleoptera is now thought to be the most species-rich group of organisms, or that there are estimated to be more species of fungi than of flowering plants? Every single one described by scientists is still assigned a binomial name and a place in the taxonomic hierarchy you were instrumental in formalizing. Several different scientific organizations, the International Association for Plant Taxonomy, the International Commission on Zoological Nomenclature, and equivalent groups for microorganisms (including fungi), now meet every few years to discuss and revise nomenclature rules. Just as in your time, decisions about nomenclature are sometimes contentious, and the need for scientists to speak clearly and unambiguously about particular groups of organisms acts as a brake on more radical proposals. I strongly agree with your comment in *Philosophia Botanica* that the names of genera are the "common currency of our botanical republic." The standardization of botanical nomenclature you spent your life promoting was and continues to be necessary to allow clear communication even among people who use different systems of classification. Although scientists today do not share the same views about the fixity of species or genera that you held when you proposed to 'fix' the contents of genera, binomial names continue to be easy to use in conversation both in person and in writing, while uninomials, long descriptive names, names made of numbers, and so on, from Albrecht von Haller to Georges-Louis Leclerc, Comte de Buffon to more recent experiments such as the PhyloCode, have never become widespread, and are most typically forgotten after only a few decades. Your leadership in the standardization of stable yet flexible nomenclature is as relevant as ever.

Your sexual system of plants, on the other hand, though very popular for nearly 100 years in the English-speaking world, has long been discarded. Scientists now focus on the natural method, as you had wished for them to do. The system we use today largely grew out of the one proposed by Bernard and Antoine-Laurent de Jussieu (remember them from your trip to Paris?), but now it is informed by the concept of evolution by descent with modification, otherwise known as natural selection, a scientific principle elaborated by the Englishman Charles Darwin in the 19th century. Further developments in the understanding of inheritance, particularly the discoveries of Mendelian ratios, the functions of chromosomes, and of the chemical structure of DNA, a molecule that is largely responsible for the form of each living thing, have long surpassed your cortex-medulla theory of the inheritance of characters in plants. We have also developed many new tools to analyze plant characters. Chemistry has advanced sufficiently so that we can now determine which molecules are responsible for plants' medicinal properties. We can use DNA analysis to see how plants are related to each other. You may be delighted to know that the water lily *Nymphaea* that you and your successor Darwin also found so puzzling is now known to come by its ambiguous features because it diverged in evolutionary history from the rest of the flowering plants before the monocots and dicots diverged. We can also use molecular biology to learn how plants function. We now know the mechanisms for how, for instance, the radially symmetrical *Peloria* flower is produced from the bilaterally symmetrical flower of *Linaria*. One thing that I know would have impressed you very much is the discovery in Mexico of *Lacandonia schismatica*, a plant that has its flower parts inside-out, that is, the stamens are in the centre, with the pistils around them! A number of botanists are now studying the molecular pathways responsible for this unusual arrangement.

The white waterlily, *Nymphaea alba* L. (Nymphaeaceae) from Linnaeus' herbarium (LINN 673.4)

The same techniques of DNA and molecular biology are also used in medicine to understand human diseases. These tools, coupled with a much better understanding of microorganisms than anyone could achieve with 18th-century microscopes, has led to a completely different classification of diseases than you had envisioned, and cures for many problems once thought incurable. We still have a long way to go, but I know you would be happy to see how far medicine has come in the 21st century.

Many advances have also been made in the way we organize information.

Index cards, like the ones you used when you were assembling material for the later editions of *Species Plantarum,* became popular in the 19th and 20th centuries. In the 20th century, electronic computers have made keeping large tables of information—what we call databases—even easier to do than before.

Another major development in the management of information about living things took place at the end of the 18th century, in France. My own research on the history of biological classifications has shown that very few naturalists in the 18th century understood your statement in *Genera Plantarum* that "natural orders instruct us in the nature of plants; artificial ones teach us to know one plant from another." The majority looked to the natural method to provide everything they wanted in a classification—not just comprehensiveness and an accurate representation of plants' relationships to each other, but also ease of use for plant identification. They held this position even though it was common knowledge that the natural method was much more difficult to learn, to teach, and, for beginners, to use for plant identification, than any artificial method, including your sexual system. You must admit, though, that, like other 18th-century botanists, you were also convinced that artificial classifications were a temporary measure that would eventually be discarded once the true natural method was discovered. This turned out not to be the case, particularly as the number of known species continued to increase exponentially throughout the 18th and 19th centuries. Just think—you had described around 7,700 species of plants in the 12th edition of *Systema Naturae* in 1767, but more than 100,000 were known a hundred years later, and the total number of species of plants on earth was then estimated at 400,000. (Today the number of described plants is thought to be around 300,000 and the total estimated number, still around 400,000.) Memorizing the descriptions of so many plants would, I think, surpass even the memory of a 'superbotanist'. You were probably cognizant even in the 1760s that looking up information about so many plants was an immense challenge, particularly to novice botanists. While artificial systems helped alleviate these problems by forcing focusing on particular features of specimens one at a time, I know you are well aware that none, including your sexual system, was capable of keeping natural genera together without breaking its own rules. Those who used your sexual system had to remember to look for palms in their own appendix, and not in the classes Triandria or Hexandria, where their flower structure would have placed them, and to look for *Valeriana rubra*, which has one stamen, with its congeners in the class for three-stamened plants, Triandria. As the number of known plants increased, memorizing exceptions such as these became more and more difficult. Botanists found ways to get around the limits of their memories by producing arrangements of plants that either had no exceptions to their artificial rules or no exceptions to the natural genera.

A French botanist, Jean-Baptiste de Monet de Lamarck, was the first to produce a fully artificial arrangement of plants of this kind. French botanical instructors active in the wake of the Revolution, such as François-Noël-Alexandre Dubois, found the *"méthode analytique"* that Lamarck introduced in his *Flore Françoise* (Flora of France, 1779) to be perfectly suited as a counterpoint and guide to the well-elaborated and largely exception-free natural method that Antoine-Laurent de Jussieu published in 1789. Dubois' work, an 1803 flora of the Orléans region, provided readers with three distinct ways to find information about plants: 1) they could analyse the plant's features one by one using the artificial component, much as users of your sexual system looked to the number and arrangement of stamens and pistils to determine plants' artificial classes and orders; 2) they could mentally compare the overall appearance of a plant to familiar plants, then turn to descriptions of the familiar plants to find the description of the plant they sought, since descriptions of similar plants were clustered together in the text, as you had clustered similar genera together in your natural orders, and, 3), if they knew the plant's name, they could look it up in the alphabetical index. Readers took to this format immediately because it combined the best features of artificial arrangements and natural arrangements and it gave them the flexibility to identify plants using whatever information they had at hand. The great success that Lamarck's successor, Augustin-Pyramus de Candolle, achieved with his 1805 edition of the *Flore Française*, which had the same format as Dubois' work, set the standard for almost all subsequent identification manuals.

Since the 19th century in particular, technologies for reproducing accurate drawings and, later, photographs, have made such great improvements that almost all biological works are illustrated. Your concern that illustrations detract from systematic works is now shown to be unfounded, particularly in zoology. You might be amused to know that the United States Fish and Wildlife Service survey of 2006 found that birdwatching is a hobby pursued by close to 50 million people in the USA alone! The popularity of this sport stems directly from the availability of well-illustrated books about birds. Illustrations also feature prominently on websites—electronic documents accessible anywhere in the world—about every possible aspect of biology. I think if you saw the kinds of materials about organisms and how to identify them available so freely and quickly on the internet today, it would strike you speechless!

Turning back to science as the professional activity it has become, systematists are still arguing (as they always have argued) about the best methods to use to represent the interrelationships of living things. In the 20th century, for example, many competing techniques were championed, notably numerical taxonomy and cladistics. You were not very impressed with the French botanist Michel Adanson's (1707–1806) proposal of a so-called natural arrangement of plants generated by the combination of different artificial ones. In the 1960s and 1970s, numerical taxonomists lauded Adanson's system as a precursor to their own, though they never claimed to produce anything other than useful artificial classifications themselves. Cladistics, roughly a technique of generating phylogenies (family trees) of organisms based on the order in which their features differentiated from those of their ancestors, is now the most accepted way to proceed, though arguments among cladists about which algorithms to use and how to cope with homologous features rage on. The latest flashpoints associated with identification and nomenclature centre on DNA barcoding and the PhyloCode. Promoters of DNA barcoding have proposed simplistic DNA tests as proxies for more in-depth systematic analyses, and have secured large research grants to do so, while proponents of the PhyloCode want to replace binomial nomenclature associated with taxonomic ranks with naming conventions built upon phylogenetic relationships, however unstable those may be. Just as with artificial systems of the 18th century, I have my doubts about whether DNA barcoding in its present form will produce results congruent with natural groups as determined by other, more comprehensive analyses. Do you see the resemblance between the uninomials-with-descriptions proposed as names in the PhyloCode and the unwieldy descriptive names botanists in the 17th century and early 18th century used before you made them obsolete?

Given the discussions about these proposals today, I think you were overly optimistic that debates of these kinds would end once the natural method was fully elaborated. I think that, no matter how much we learn about the natural world, these debates will always continue in some form. Not every scientist understands or will ever understand the necessity you pointed out of using separate techniques for describing plants' relationships to each other and the techniques for identifying them. Not everyone understands, as you do, the inherently practical nature of taxonomy. For these reasons, I

Laur Alstrin's pen and ink drawing of Sjupp, Linnaeus' pet raccoon

can't see the disappearance of the binomial nomenclature that still bears your name any time soon. It's far too useful, conforming as it does to the general linguistic tendency to use nouns with adjectives to describe things, and how easy it is to remember names that follow this pattern.

So, rest assured, Professor Linnaeus, just because you have not published anything since the late 1780s does not mean that you have been forgotten!

One more thing. It is a warm June evening here in Toronto. I am writing this from my apartment. My pet fish, *Abramites hypselonotus*—a freshwater species from Venezuela (first described in 1868) are munching on the leaves of plantain (*Plantago* sp.) and dandelion (*Taraxacum* sp.), weeds indigenous to Europe but naturalized all over North America. Outside my window, skunks (*Mephitis mephitis*) and raccoons (*Procyon lotor*) are fighting over fallen mulberries (*Morus* sp.). (You may miss your pet raccoon, Sjupp, but then again, he was never provoking skunks to spray in your vicinity, now, was he?) The trees are full of birds attracted to the berries—indigenous species such as robins (*Turdus migratorius*), mourning doves (*Zenaida macroura*), and northern flickers (*Colaptes auratus*), and also European species, such as starlings (*Sturnus vulgaris*) and house sparrows (*Passer domesticus*). Sweden, like Canada, now likewise hosts a mixture of indigenous and introduced species. These introduced species, coupled with, most importantly, the destruction of habitat, have led to many species becoming extinct. This means that these species have completely died out everywhere on earth. Naturalists when you were in your prime never discussed extinction; even when they began to talk about it at the beginning of the 19th century, it was a controversial topic. Now, the 'biodiversity crisis', or rapid loss of species, is a prime concern to everyone interested in living things. You might feel very uncomfortable in this modified environment. As much as I and other naturalists enjoy observing the natural world, our enjoyment is tinged with sadness and fear about the threat to the diversity of life on earth. Systematists are now under more pressure than ever before to understand as much as they can about biodiversity, before more species disappear.

The jacket and shirt of 'Linnaean' clothing at the Linnean Society of London

The acknowledgement of the biodiversity crisis has had one effect of which you will likely approve, though. Over the last 150 years, systematics had gone from being a highly respected science that other disciplines, such as chemistry and mineralogy, looked to for inspiration, to one of the lower-status scientific pursuits, eclipsed in the 19th and 20th centuries by chemistry and physics. Even within the life sciences, physiology, biochemistry, genetics and other methods for understanding the internal functioning of organisms, rather than their relationships to each other, have eroded the influence of systematics. The loss of biodiversity now facing us is bringing systematics and the importance of systematists' expertise back into the spotlight, though systematics still attracts less funding than many other branches of biology.

All this may be difficult for you to absorb; it *is* a lot of things to think about for someone who has been out of the loop for so long. However, rest assured that many scientists would be happy to bring you up to date. And you will always have a place to stay in Uppsala—Uppsala University there has preserved your house and garden at Hammarby just as you left them, while your herbarium, books, and manuscripts are safely stored at the Linnean Society of London. There is even a set of clothing for you hanging in a back room at The Linnean Society library, should you decide to drop by. I think you would give the staff and visitors there the surprise of their lives.

With warm regards,

Sincerely,

Sara Scharf

Sara Scharf, FLS (Fellow of The Linnean Society of London)

Magister

I shall call you that unashamedly, for even more than three hundred years after your birth you stand Colossus-like over our discipline. There has been a common stream of endeavour running through taxonomy since your time. Why are those of us in the field inclined to glance back, nay more than glance, when science almost defines itself by looking forward—forever pushing the boundaries? Partly it is because the system you gave us for naming species has given us nomenclatural stability. Priority has been a guiding standard for stability since the taxonomic community adopted your works of 1753 and 1758 as starting points for naming plants and (nearly all) animals. Yet that is not the whole story, for otherwise why have so many written about your life and why do they continue to do so? It is not just about the way you articulated and established the binomial system, but that by so doing you facilitated the way information about animals, plants and even minerals was exchanged. In your day, the community that shared this information was much smaller than now; but your system has scaled up and lasted. You led the way. Yet you were not just an armchair systems man, but a dedicated and practical taxonomist, cataloguing and describing species. You travelled in your early years, although not outside northern Europe, and even contributed to the genre trivialised today by the term 'travel writing'. But after your early wanderings you seemed happy to settle down, travelling vicariously through your students. Many of these disciples roamed widely, including to tropical and subtropical parts of the globe. Several never returned. Nearly all, however, sent you specimens that had been collected mainly from or around the ports of the old shipping lanes of the world where their vessels berthed, and that gave you a global view of nature.

I think we admire your vision and its translation into material outcomes. Forgive the language—I try to stop myself, but we are all children of our time and linguistic fashions are insidious. You remained more pure, sticking to Latin for your scientific writings. Today Latin is the universal language that no one understands (botanists try), but in your day it was the means of academic communication across language barriers. They say your capacity for writing in Latin was functional and at times rather inelegant. Perhaps you fell down a bit in your application of grammar. But your names were often subtle, as was your appreciation of classical mythology: who else would choose the name *palaeno* for a butterfly that, like the Nymph, played and danced on the moors? Science then had no boundaries to cross although it was emerging as a rational exercise. Today we have become so focused that we are rediscovering the need to be inter-disciplinary. You were never troubled by recreating

The Red Admiral, *Vanessa atalanta* (L.) was described as *Papilio atalanta* by Linnaeus and is named for the Greek mythological character Atalanta. Atalanta was fleet of foot and promised her hand in marriage to any man who could pass her in a race, but death to any who lost.

The labels from the specimen of *Vanessa atalanta* in the Linnaean collections; the lectotype was designated by Scoble & Honey in *Zoological Journal of the Linnean Society* 132: 302 (2001), with the other label "119 Atalanta" in Linnaeus' hand.

an 'arts/science interface' (so never suffered its pseudo-intellectualism).

I wonder what you would think of taxonomy today. You would surely recognise many similarities—establishing concepts of species, giving them names and linking observations to the right names. You would see also many changes, gentle Uppsala would doubtless be a shock . . . and all that habitat change even in eco-friendly Sweden. Two things, however, might seem enduring—the species you described would look much the same (so I wonder what you would make of Mr Charles Darwin). And human nature—that has changed little enough.

But back to the vision: a taxonomy of all living things—that's what you were after. By contrast, your English contemporaries such as Gilbert White of Hampshire and Henry Seymer of Dorset (both great men too) were content to focus largely on the detail of their local areas. Well, we have found rather more species that you might have expected—even if most of them seem doomed soon to extinction. But you went for the 'big picture' (sorry again)—and what a canvas. How proud you would be to know that so many of your names endure in our present day classifications.

Part of me remains curious at the measure of the esteem you continue to enjoy, for we are moved today more by the dynamic than the static—process over pattern, Darwin over Linnaeus. So let me inject a modern thought. Some of us think taxonomy is, at base, a system of information—and managing information is a modern obsession, and a fearsome task. It's not just taxon names, but all the data associated with those names. We have means now, of which you will know nothing, for helping us cope with and exchange voluminous amounts of information. We also have methods for deep analysis of those data and techniques by which they can be portrayed graphically. So your system and the way you associated information about species with names was an early attempt to forge what we now call an information science. Our modern systems change frequently: your system remains with us because it works and includes those key and lasting standards of binomial nomenclature and priority in naming.

Of course there were other elements to your life—your practical enthusiasm for the Germanic economic system of cameralism, by which national prosperity was a proper part of national governance (even if you did not worry too much about the effect on those countries from whence the natural products to support it came). Trying to grow tea in Uppsala was a measure of that enthusiasm, even if the Swedish winter dealt the experiment a mortal blow. Yet the thought was worthy—your country, after all, was desperately poor so growing things to consume was an appropriate response. Linking scholarship with social value—I applaud the sentiment; we should do more of it today, particularly in taxonomy.

What I admire about you is what you did for natural history—combining observations of the world around you with scholarship and systematisation. You gave us durable functionality. And we still have your collections and manuscripts—imagine that! Overall your specimens are in remarkably good shape considering their age and the state in which they were stored immediately after your demise. You would be amazed at how they are housed today, in conditions superior probably to any natural history specimens in the world—a sign of veneration of you through them perhaps. But it's not just that: we use them still, and many are the name-bearing types (you didn't use the term, though you did abide more or less by the concept) of many of our common European species. And when I worked on your butterflies a few years ago, there was that special pleasure of trying to understand your species concepts by looking at the very same specimens on which

Tea (*Camellia sinensis* (L.) Kuntze, Theaceae) growing in Lushan, China

you based those concepts. Interpreting your thoughts compressed the years between us; that common stream runs yet.

Malcolm Scoble

Malcolm J. Scoble

Dear archiater, Professor Linné,

Please excuse me for addressing you so frankly across the centuries. I am of course fully aware that you never held Denmark in high esteem and only visited my country once, on your way to Holland in 1735. After having crossed the short strait between Helsingborg and Elsinore, separating Sweden from Denmark, you and your travel companion, Claes Sohlberg, stayed at an inn in Elsinore for five days waiting for a boat bound for Holland. In the end you had to accept passage on a boat to Travemünde instead. A coincidence, that gave you an unexpected chance to visit Hamburg and study the famous, stuffed seven-headed hydra, a drawing of which was even published by Albertus Seba, and which you—of course—immediately declared a fake.

Fortunately, you admitted several Danish scholars to study under your guidance at Uppsala. Most notably among them the entomologist Johann Christian Fabricius (1745–1808) and the botanist Martin Vahl (1749–1804) who spent two and five years under your guidance, respectively. Thus, your influence on the progress of natural history at once stretched into Denmark, not least in entomology and botany, and can hardly be overestimated. I am confident, that Fabricius' books *Systema Entomologiae* and *Philosophia Entomologia* published in 1775 and 1776, respectively, surely would have pleased you, but I am afraid your failing health did not allow this.

However, I am writing to you in a different context. The Linnaean project, if there ever was one, was on the surface deceptively simple, to divide and name the natural world, though its philosophical Aristotelian-Thomistic foundation (which need not concern us here) was rather complex. *Systema Naturae* may indeed be viewed as the start of this truly encyclopedic effort, an effort which, in the twilight of your life, turned out to be no less than a formidable achievement. The creation of order out of chaos was a feat only obtainable by an extremely orderly mind—a mind bordering on the autistic.

Being a botanist myself, I will comment on your botanical systems, even though I do believe that some of my remarks have wider applicability. Of the three complete botanical systems you published during your lifetime the sexual system clearly stands out as having the greatest, both contemporary and long term, impact. Its virtue being that it immediately provided botanists (and other users of botanical information) with a badly needed, simple, easily comprehended scaffold for organizing, storing, and retrieving information. Adding to its almost instantaneous success was—of course—your introduction of an endlessly adaptable and flexible way of naming plants.

Presupposing the Aristotelian essence of plants expressed in the fructification you used two parts of it: stamens and carpels; by means of stamens (*Systema a staminibus quod Nostrum est*); and not only their number, but also their relative length, their distinctness or fusion and their presence or absence in certain flowers, you classified the plant kingdom into 24 classes, and further subdivided these classes into orders on the basis of the number of carpels. Present day botanists frequently consider the sexual system your only system and equate it with the Linnaean System. Thus, one of the most interesting facts about the sexual system, that it was intended as a means to an end and not as a goal in its own right, is neglected. I am afraid it would come as a shock to a pious man as you, Professor, that today most people have totally forgotten that you definitely considered the sexual system artificial and only a method to help us unveil the Creator's plan. That you created a system intended for beginners (*Methodus á calycis speciebus*) and struggled with a natural system (*Fragmenta Methodi naturalis*) largely goes unnoticed.

Siegesbeckia occidentalis L. in Linnaeus' herbarium (LINN 1018.2), now known as *Verbesina occidentalis* (L) Walter (Asteraceae)

The sexual system was of course not accepted right away and in a modern perspective some of the objections raised, most notably the ones by J. G. Siegesbeck, stand out as being rather amusing: "What man would believe that God would introduce such harlotry into the propagation of the plant kingdom? Who would instruct youth in so voluptuous a system without scandal?" Today, Siegesbeck is largely remembered in the name *Siegesbeckia* L., a genus of ill-smelling, small-flowered weeds. Other critical voices, in particular those raised by several French botanists, reflected a much more fundamental disagreement.

As you stressed, the intention of the sexual system was to reconcile its artificial groups with those created by God, and whereas species and genera were unchangeable and therefore recognizable by man, orders and classes were the "work of nature and art". The more plants we know, the more likely it is that the borders between the orders and classes will disappear (*Natura non facit saltus*). During most of your lifetime you struggled, sometimes optimistically, at other times pessimistically, to find the "natural method", viz. the method that would disclose the natural groups—a very problematic fact, which you openly admitted—even to the extent that you sometimes consider your contribution towards the "natural method" your masterpiece at other times as unsolvable as *quadratura circuli*.

Your insight that the 'true' natural relationships between plants were reaching out in all directions, and could be reflected as a geographical map (in *Philosophia Botanica*) is often forgotten, but proved to be a fertile and surprisingly modern idea. It was a radical break with the traditional linear ordering of organisms as depicted in the *scala naturae*—the great chain of being. To resolve the conflict between the multidimensional character space and two-dimensional genealogy remained a challenge for nearly 200 years. The map analogy, first visualized by your pupil Paul Giseke in 1792 as *Tabula genealogico-geographica affinitatum plantarum secundum ordines naturalis Linnaei*, was taken up by other botanists such as Antoine-Laurent de Jussieu, Charles Louis L'Héritier de Brutelle, Louis-Marie Aubert du Petit Thouars, and Augustin Pyramus de Candolle, and more recently by August Adriaan Pulle, and in many ways it climaxed in the works of your fellow countryman, Rolf M.T. Dahlgren (1932–1987), who was professor at the Botanical Museum at the University of Copenhagen.

Two of your innovations have survived until this date and are still in use: your nomenclature and your sexual system.

The strength of your nomenclatural system is undoubtedly that it is basically free of theoretical underpinning, and is eminently suited to store what we already know and flexible enough to accommodate what we learn. It has additionally proven its worth as both an efficient means of communication and information storage across centuries.

The sexual system is, of course, not our current botanical reference system. Too much has happened since your time and we are using analytical methods and types of data you could naturally never have envisioned. However, the sexual system remains a very useful method, now as then, for organizing information in artificial keys, and innumerable floras still use the numbers and attributes of the stamens and the number of carpels in entry keys, bearing testimony to the efficiency and ease of use of the sexual organs of plants. However, as you already noted some of the natural groups that were and still are generally accepted, e.g., species of the genus *Valeriana*, were scattered among different classes in the sexual system, as some species have one, some two, and some three stamens—some are even dioecious.

In hindsight the Linnaean project, your scientific master plan, must have taken shape at an early age—at least years before you went to Holland. Hence, the first written remnants of the sexual system can be traced back to your manuscript *Praeludia sponsaliorum Plantarum* from 1729, written during your early, and far from prosperous days in Uppsala, but there is every reason to believe that Dr. Johan Rothman, your physics teacher at the Gymnasium in Växjö drew your attention to the sexuality of plants as early as 1724. The sheer number and size of the manuscripts you brought with you to Holland at an age of only 28, is no less than astonishing and also clearly point to a young man with a definite goal in life.

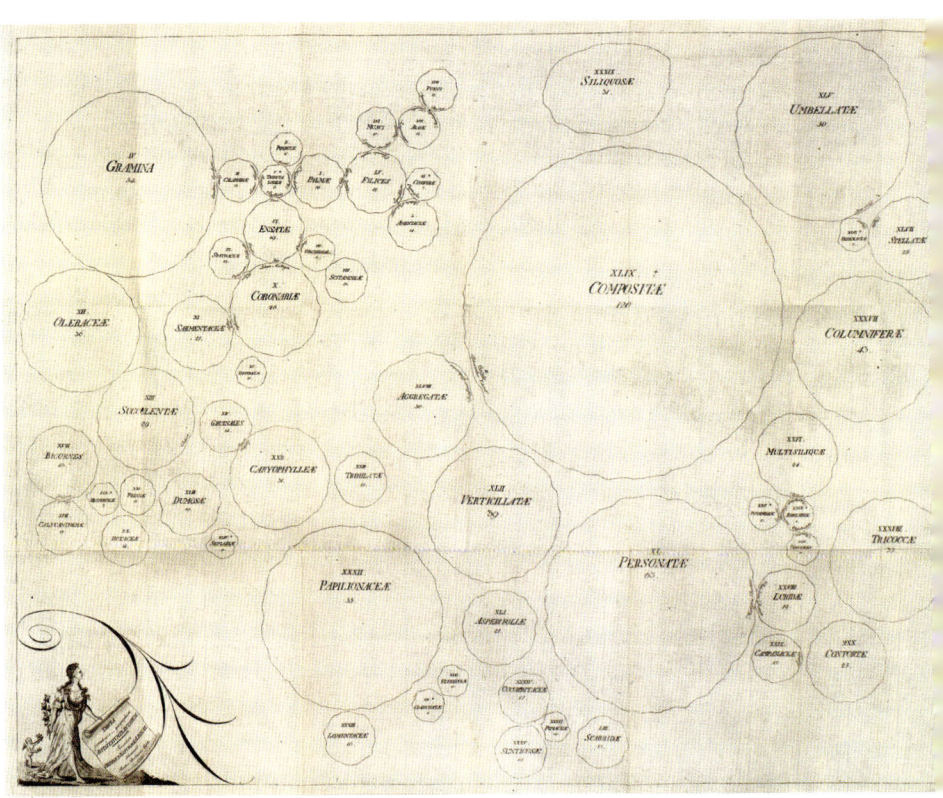

Linnaeus' map analogy as interpreted by Paul Dietrich Giseke and published in *Praelectiones in Ordines Naturales Plantarum*, 1792

It is my judgment that your own and other contemporary botanists' inability to capture the natural groups was a highly frustrating experience. I am sure that you would be very surprised to know that despite 250 years of trying we cannot help you. If you came to my office tomorrow and wanted an explanation, I fear that you would be surprised to learn that God has no role in modern biology. Perhaps you would be equally surprised to learn that essentialism is largely off the agenda, that species are not stable but constantly changing, that higher taxa are historical entities, that we are still struggling with finding the pattern in our characters, and that the tree of life is unknowable and only a hypothesis. Hence, some of the patterns you found remain—some may even remain for ever, but the process explaining the observed patterns is very different. Naturally, you would invoke a Deity, but we call the process evolution, due to radical changes in our perception of the reasons and consequences of the variability we observe, and of man's own place in nature.

I beg to remain, Sir, your most humble and obedient servant,

Ole Seberg

Carl Linnæus

Viro Domino Carolo Linnaeus, Svecus,
Professori Upsaliensis,
Botanico Consummatissimo
M. D. et Socius Academiae Imperialis et Regiae Parisinae

Salutem Dicit
Jorge Soberon

Dear Professor von Linné,

I have been asked to send you a letter, although for what purpose I cannot possibly fathom, since it is unlikely that this will reach you. Still, good friends have asked, so I am happy to comply. Allow me first to introduce myself: I was, until very recently, a civil servant in a country that in your century was a part of the then vast Spanish Empire, the first one in which the sun never set. About 15 years ago, I was given a task that involved your ideas but in ways that perhaps you did not foresee. I ask humbly that you have the patience to hear about this. Of course, you were very much aware of the Vice Royalty of New Spain, since you received specimens and material from that part of the world. Many names of plants and animals of my country still bear the

The Cardboard Palm *Zamia furfuracea* L.f. in Aiton (Zamiaceae) is native to Mexico and is a cycad, not a palm

evocative 'L.' as a proof of the endurance and wisdom of the logical scheme of the *Systema Naturae*. I can imagine you, in your cabinet in Uppsala, receiving a steady flow of specimens from remote parts, perhaps trying to picture them in your mind, as some sort of distorted version of your Swedish (or even worse, the Low Countries!) meadows and forests. Were you intrigued by those huge beetles, like *Coelosis biloba* L. or *Cyclocephala amazona* L.? Or by that cycad, *Zamia furfuracea*, that perhaps you examined and your son misplaced in the genus *Palma*? What kind of landscape could grow these strange species?

You worked, perhaps fortunately, at the beginning of the golden age of exploration. Your publications described about 12,000 species, a fairly manageable number. You did not know about the existence of so called 'megadiverse countries' (an unfortunate neologism coined by a friend of mine that would have caused horror to someone with your classical education), in which 12,000 species can be found in a few hectares of land. Did you, in some nightmarish dream perhaps caused by ingestion of a bit of *Amanita*, even consider the extent of the task that your successors would face? I do not think so. The truth is far richer and weirder than any fiction and without experiencing it firsthand, how could a Swede of the 18th century imagine the unbelievable wealth of forms, colors, landscapes, smells, and natural histories to be found in a tropical forest, in a coral reef, or in the so-ineptly called 'deserts' of the New World? That your *Systema* would have to cope with thousands more species those you described on a lifetime of hard work?

Amanita muscaria (L.) Lam. (Amanitaceae), the Fly Agaric, from Sarah Price's *Illustrations of Fungi* (1864)

Which brings me back to the task that a former 'president' (some sort of King) of Mexico gave to his natural philosophers. The last truly autocratic president of my country, a man of culture and vision, decided that the country needed a proper inventory of its "biodiversity" (another neologism of both dubious utility and distinct ugliness). Without such an inventory the proper task of governing the *Res Publica* could not succeed. This President was a man living in the wrong age. Perhaps he truly belonged to your own time. But I digress. He listened to an important Mexican *savant*, was convinced by him, and then he simply gave the order: "perform the inventory of the biodiversity of Mexico, and maintain it updated". And then, after descending several steps in the echelons of governmental hierarchy, the order ended on my lap, and I had to start working on this colossal task. My dear Linnaeus! I was so happy that you had invented your *Systema*! You know, to organize the inventory, we needed names, arranged in some logical structure. Your hierarchical system, and your Latin binomials, so dreaded by biology students in my day, were simply indispensable to fulfill our orders. Today, you know, we use machines called computers (which you would love): the entire logic of your system can be represented in abstract codes that these machines can 'understand,' in some arcane way. Your names then open the door for these machines to search, for example, for medicinal plants (wouldn't this had been useful to you?), or for those species that we do commerce with, or for the ones that we wish to protect. These machines now organize about 90,000 names (just for my corner of the New Spain) and data about more than 3 million specimens, with almost 100,000 images of high quality (and I am not sure how many documents) and this is all built around your concepts.

It is amazing, is it not, my dear Linnaeus, the power of a good idea. Of course, it is not surprising that naturalists all over the world now use your system. But civil servants in government offices, with machines unimaginable in your age? Mighty good idea, this *Systema* of yours, as they say in the part of Louisiana in which I now live. Well, I wanted to tell you about this because, if somehow this letter would reach you (the organizers have promised to try), I would very much like you to receive the most sincere thanks from a civil servant that could not possibly have fulfilled his orders without the support of your great idea and hard work.

Vale, Vir illustrisimo. Ars vestra Decus & Gloria
Dabam ex museum Kansiensis
Anno Domini 30 *Iulii* 2008

Carl Linnæus

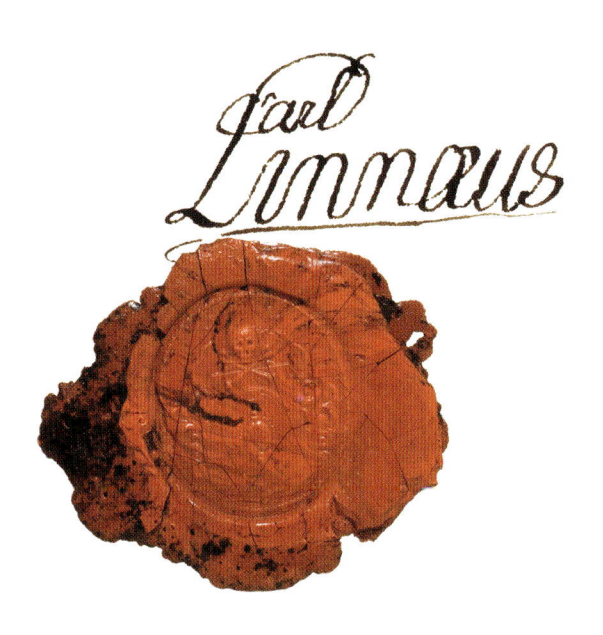

Daniel Solander (1733–1782) was perhaps the most celebrated of Linnaeus' pupils; he accompanied Joseph Banks on voyage of the *Endeavour* under the command of Captain Cook. This short note was sent from London and does not bear a date, but is probably from after Solander's return from Australia.

"Please find enclosed the description of the insects which I sent in the box. I forgot to put it in. When you are sending me letters [that are] addressed directly to me I am most sure to get them if you address them to Brander's and Spalding's in this way: To Mr Solander – to be delivered at Mr. Spalding & Brander in White Lyon Court, Cornhill, at London. If the house address is not put on the letters, one may often run the risk of losing the letters. . . ."

To
The Royal Society
of Sciences,
at Upsal.

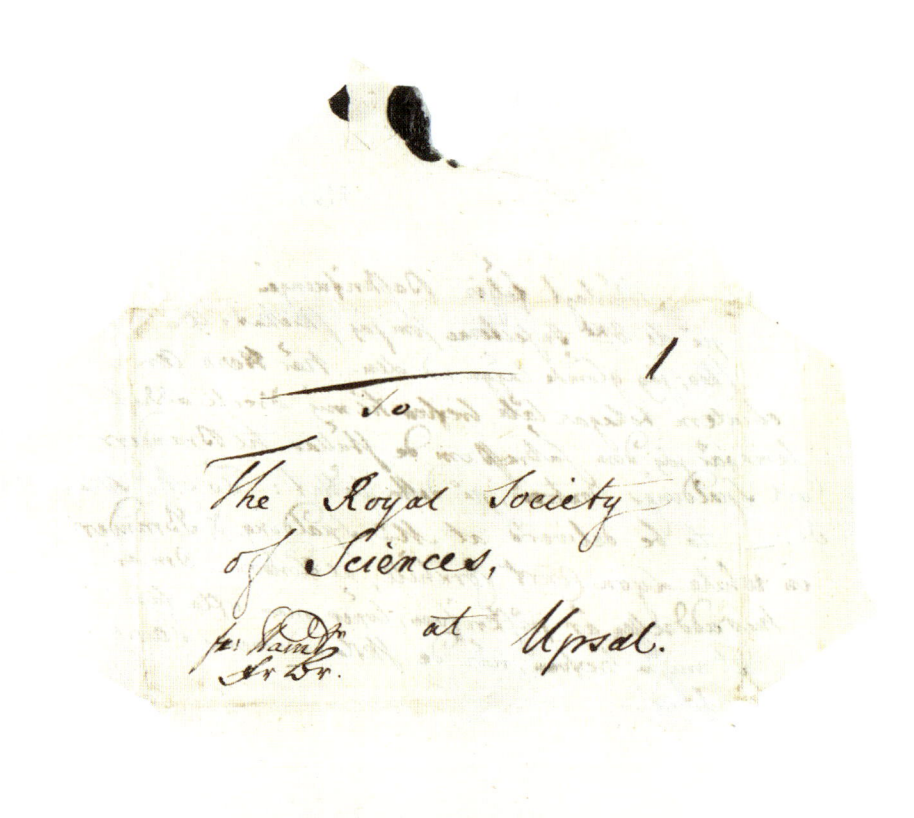

THE NEW YORK BOTANICAL GARDEN

July 24, 2008

Dr. Carl Linnaeus (Baron Carl von Linné)
Professor of Botany and Medicine
Uppsala University
Uppsala, Sweden

Dear Professor Linnaeus:

I am writing this letter as an evaluation of your nomination for Professor Emeritus. I write this letter as an evaluation of your research and teaching career by an external peer–review committee. I base my comments on having followed the conclusions of the committee as summarized below, as well as on my own evaluation.

With respect to your scientific productivity in terms of publications, while the quantity is truly impressive, the committee noted that it is dominated by single-authored works. Moreover, these works do not appear to be peer-reviewed and do not appear in journals with high impact factors so the future use of this information could be quite limited. They also noted that you have a tendency to publish your student's works under your own name, something not really acceptable today, but perhaps a presage of the future.

The review committee expressed concern that your publications were basically descriptive, applied, and written in classical Latin, a dying if not dead language. They were unable to discern a theoretical framework for your work. Moreover, although you have collected and published a vast quantity of empirical observations and data, your work is seriously lacking in statistical methods, approaches, and the significance and support values so necessary to validate both the observations and the conclusions.

The committee was impressed with your teaching skills and the outstanding evaluations given by the undergraduate student body over the years. It did note, however, that many of your graduate students had a tendency to travel widely, spending many years away on long journeys collecting plants for you that you subsequently published. It also noted that many failed to return from these journeys. As a result, the committee expressed concern that you have communication and personal problems with advanced students and colleagues and that

Some of the many dissertations of Linnaeus' students housed in the Linnean Society of London library

this could be deleterious to relations with the faculty and the overall well-being of the emeritus faculty.

Research funding was another area that seemed problematic to the review committee. They felt that there was a paucity of peer-reviewed funding and too much non-competitive funding via the royal family and other private sources, which may have biased the direction of the research and publications.

The review committee concluded that the problems outlined above were most likely the result of a lack of research focus and failure to take a more reductionist approach. As examples, they cited work on fish, insects, mammals, geology, minerals, ethnobotany, etc. It appeared to them that you are a jack of all clades and master of none. They also pointed out that your interest in botany bordered on the prurient. With respect to ethnobotany, your efforts were considered superficial in that you failed to follow through with a study of the efficacy of medicinal plants even though you were a practicing physician. It was felt that you spent too much time in reconstructing native games and in dressing in native garb that was unbecoming of a university professor, particularly one with royal support. Additionally, too much potentially productive time was spent in horticultural pursuits in developing a botanical garden and pursuing the development of cultured pearls. The committee did express satisfaction with your having helped Celsius develop the centigrade thermometer in its current form and scale.

As a result of all of the foregoing, the review committee concluded that it could not support your nomination for Professor Emeritus.

I have reviewed very carefully the comments and conclusion of the review committee as well as the materials supplied for your nomination for Professor Emeritus. In my opinion, the review committee was too focused on details, numerical considerations, and so-called cutting edge science. In short, they failed to see the forest for the trees because they themselves took a reductionist approach in their evaluation. Your career accomplishments, put in a larger context, are most impressive. Your research is not just prodigious, but is also diligent. It has clearly led to a better understanding of the world around us while at the same time provided us with a way of organizing the vast amount of information we have about the natural world. Therefore, my assessment is not congruent with that of the committee. I am very pleased to be able to inform you that your nomination for Professor Emeritus is approved—I congratulate you on your accomplishments and look forward to many more years of service to the academic community and society at large.

Sincerely,

Dennis Wm. Stevenson, FMLS
Vice President & Pfizer Curator for Botanical Science

Carl Linnæus

27 February 2008

Professor Carl Linnaeus
Botanical Garden
Uppsala University
Uppsala, Sweden

Dear Professor Linnaeus:

As you may or may not be aware, many modern plant systematists view our field primarily as tubes filled with DNA extracts and branching diagrams based on quantitative phylogenetic algorithms. To them you are only a distant figure, of historical importance, but of no particular modern relevance. For me, this has not at all been the case. I have drawn inspiration from your books, your scientific contributions, and your human qualities.

As a confirmed bibliophile, I could hardly avoid seeking copies of your many books, although due to limited personal resources, only by acquiring facsimile editions. From your first book, *Systema Naturae* (1735) on beyond to the *Species Plantarum* (1753), it is noteworthy how you observed, edited, and synthesized information. Particularly impressive were your industry, rapid development as a scientist, and productivity during years in Leiden, 1735–1738, between the young ages of 28 and 31. Your employ with George Clifford at the country estate of Hartecamp outside of Leiden offered you the resources to work unfettered by mundane tasks. This period resulted in publication of *Musa Cliffortiana* (1736), *Fundamenta Botanica* (1736), *Bibliotheca Botanica* (1736), *Critica Botanica* (1737), *Hortus Cliffortianus* (1737), *Flora Lapponica* (1737), and the first edition of *Genera Plantarum* (1737). This was an amazing set of publications from a young scientist. When I look at this immense output, I gain strength for moving forward with my own work. It is perhaps no accident that the part of my personal library that contains these books lies conspicuously to the right of my office desk.

Your scientific contributions are impressive in that they were totally appropriate for the time. You believed that you had a special calling to chronicle much of the natural world, as well as later in life, to contribute to the improvement of the Swedish economy, and chronicle you did—listing nearly all aspects of botany (plants, literature, morphological features), plus much on animals, nutrition, etc. There is hardly any aspect of natural history that wasn't organized by you into a system of some kind. This passion for order was exactly what was needed in the 18th century. This was an age of discovery of new organisms and

Frontispiece of *Flora Lapponica* (1737) by Carolus Linnaeus

information about them, and comprehensive systems were needed so that these new data could be stored, retrieved, and evaluated. It has been often quipped that: "God created, but Linnaeus ordered." More importantly, you ordered at *the right time*, and in the process strongly accelerated development of the botanical sciences.

Your personality to me has also been inspirational. Yes, it has been said that you were self-promoting and arrogant. I mean no disrespect when I remark that you sought friends in high places and used these contacts to further your career. But, to a certain degree these qualities are not objectionable. They are, in fact, almost essential for scientific and academic success, even today. But beyond these qualities you were a teacher, who relished having students, and who loved lecturing to them about aspects of "sweet flora." You also wanted students to enjoy themselves and have fun, as was evidenced by the numerous parties you organized for them.

You may be amused, and I hope flattered, to learn that each year I put on a tailor-made costume from the 18th century, including wig and buckle-shoes, and return to the general botany classroom as yourself, having arisen for the purpose from your resting place in Uppsala Cathedral. There you stand recreated as an old man recounting your life, giving your achievements, lauding the successes of your students, and sharing some of the joys and sorrows during your own time on Earth. This brings Linnaeus alive for the students—and they never forget you. During the hour I also lose myself in the character and have the honor of communing directly with you. You were totally dedicated to a life of science and education, and even by modern standards you continue as a successful role model for the present and future generations of systematic botanists.

Many thanks for all you have contributed to our field. We owe you much.

Most sincerely,

Tod F. Stuessy
Professor and Head
Department of Systematic and Evolutionary Botany
University of Vienna
Vienna, Austria

Carl Linnæus

A letter to Carolus Linnaeus

As a young student beginning the study of Zoology, among the first lectures that I heard was a most significant one—an introduction to the concept of a system of naming organisms based on a binomial nomenclature. It became impressed on me that the 10th edition of your major work, the *Systema Naturae*, was the starting point for all zoological names, as this was the first edition in which you consistently used binomials to describe animals.

The timelines of history are such that the era in which you produced your *magnum opus* was the century in which my Huguenot ancestors fled Europe for the Americas. After two hundred years the world is still beset with religious intolerances but science has traveled far and at an ever-accelerating pace.

As my studies progressed I became interested in small flies, the Chironomid midges. When I was first introduced to one of the seminal works of the 20th century, a monograph by Henry K. Townes, Jr. entitled, *The Nearctic Species of Tendipedini [Diptera, Tendipedidae(=Chironomidae)]*, published by *The American Midland Naturalist*, Vol. 34, 1945, I was amazed to see a species to which you had first given the binomen, *Tipula plumosa*, in 1758.

Chironoma plumosa (L.) (Tipulidae), the midge originally described as *Tipula plumosa* by Linnaeus

Linnaeus' label on his specimen of *Tipula plumosa*

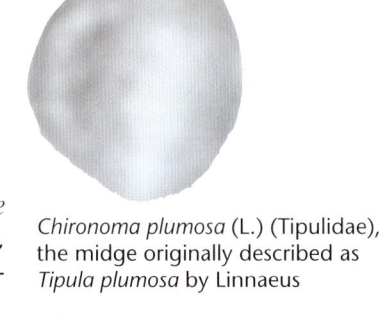

We now know that many species with high vagility, such as small midges

and gnats, become components of an aerial plankton and are transported for great distances. Thus, your species, *Chironomus plumosus* (Linnaeus) (as it is now called) is circumpolar in the northern hemisphere.

The genius of your contribution in using a noun and a single modifier to designate a species cannot be overemphasized. If one understands the etymology of the two names they have great mnemonic value as an aid to memory.

Many scientists have used the naming of new species in dedication to someone or to recognize the locality of origin. Family members are a frequent source of names and as a taxonomist I'm to be included in that category.

- I have recognized my late wife's contributions with the names *Paramerina smithae* (My wife's maiden name, Smith); *Maryella reducta* (my wife's given name + a brachypterous Subantarctic midge);

- My youngest daughter, Amy, with the anagram, *Yama tahetiensis* for a midge from the buttress pools of a tree in Taheti;

- *Cladotanytarsus marki*, named for my youngest son, Mark, for a midge from the Grand Canyon.

One of my favorite midge names is *Dicrotendipes thanatogratis* Epler, dedicated to *The Grateful Dead*, a musical group that John Epler, the author of the species name, found so inspiring in his youth.

In this era of rapidly advancing science the taxonomy of biota is seen by some as on a par with collecting buttons or other memorabilia. Not so! Before one can use the techniques of DNA analysis an initial morphological separation is necessary. Among the many mosquitoes swarming around a victim only one or two species are of concern as disease vectors. Morphological taxonomy is needed before the refinements of DNA can be employed.

It has been argued that the binomial system is outdated and a more reasonable alphanumeric system should be utilized. Some efficiency might be gained in computer usage but the mnemonic value of the binomen would be lost. Science is cosmopolitan and scientists studying midges are found in most countries, each speaking different languages. The universality of the latinized binomen is the commonality binding all colleagues into working towards a common goal.

My congratulations to you for having the insight that an abbreviate latinized phrase, the scientific name, is a basic descriptor for all biota; it has well withstood the test of time.

With continued admiration,

James E. Sublette

James E. Sublette, Ph. D. (Zoology)

Professor Emeritus
Colorado State University-Pueblo
(University of Southern Colorado)
Formerly, Professor of Biology and
Distinguished Research Professor
Dean of Graduate Studies and Research
Eastern New Mexico University, Portales.

Carl Linnaeus

Dear Carolus,

Years ago (1761) you wrote your friend, Baron Nikolaus von Jacquin, "while my colleagues daily enjoy the pleasures of this existence [as professors at the University of Uppsala] I spend days and nights in the exploration of a field of learning that thousands of them will not suffice to bring to completion, not to mention that every day I have to squander time on correspondence with various scholars . . ." [translated quote from Blunt (1972, *Linnaeus. The Compleat Naturalist*, Princeton University Press, page 168)].

A selection of Linnaeus' own wax seals

So today I write to you about where we are some 250 years after the 10th edition of *Systema Naturae* was published. That work today is accepted as the start of modern systematic zoology. Yes, you did provide two more editions and a German worker did provide a 13th. So, my references are in respects to that 10th edition.

As you foresaw, the task was much greater than one individual could complete. So, your students divided the task among themselves. Johann Christian Fabricius took over the task of the Insecta and produced a series of Systemae, but in the end of his career, he, too, recognized that the task of summarizing even all insects in one work was too much. So, he began to summarize knowledge by orders. Hence, his work, *Systema Antliatorum* (1805) (he abandoned the Aristolean system that you followed based on wings for one based on mouthparts, so his Antliata is the same as your Diptera). While you knew 191 species of flies distributed among 10 genera, Fabricius summarized information about 1,151 species in 78 genera. His was the last treatise to summarize all knowledge about flies. Subsequently workers divided that task either geographically or taxonomically. A Hungarian, Kalman Kertész, attempted to make a comprehensive index to all the work on flies with his catalog (1902–1910), but abandoned the task as impossible after covering less than half of them.

Today, we are again attempting, with our access to new technologies, to do what you and Fabricius did and Kertész attempted to do. We want to summarize all that is known about flies in a medium accessible to all everywhere. The first step is to accumulate and evaluate all the scientific names used for flies. This we are attempting to do with our Biosystematic Database of World Diptera (go to www.diptera.org). So far we have information on 156,646 species distributed among 11,698 genera in 236 families.

Next we want to have a summary statement for each species, having minimally an image (the type specimen or another voucher); distribution map showing at least one occurrence (type-locality); and linkage to what has been published on the species. This is your *Systema Naturae*, but today we call it the *Encyclopedia of Life* as generated by the Global Biodiversity Information Facility and a consortium of six other organizations, led by the Smithsonian Institution.

Temnostoma vespiforme (L.) (Syrphidae) – this flower fly was first described by Linnaeus as *Musca vespiformis*

Labels from the type specimen of *Temnostoma vespiforme* in the Linnaean collection

The success of this new Encyclopedia of Life depends on public support, which unfortunately has been minimal.

Beyond the challenge of summarizing what is ALREADY known, we need to make known what is unknown. Like yourself, we today suspect that we know only about 10% of the life that really exists on earth. We, today, have better means of collecting samples of Nature everywhere and have better means of evaluating these samples. As in your day, we can look at these samples and assess similarities and differences in appearances, but we can also look at characteristics hidden from human eyes, the DNA sequences, which are a representation of the basis for all inherited characters. However, regardless of the new wealth of information about organisms, from morphology to DNA, specialists like yourself are still essential for the proper evaluation of diversity and to create a meaningful system to summarize our natural world.

So, today, we remain exactly where you were when you wrote to Nikolaus. We know what the task is, how large the task is, but we remain unable to find the resources to accomplish the task. And we, human society, remain diminished because of that.

Sincerely,

F. Christian Thompson
Systematic Entomology Laboratory
Agriculture Research, USDA
Washington, DC 20560

Carl Linnæus

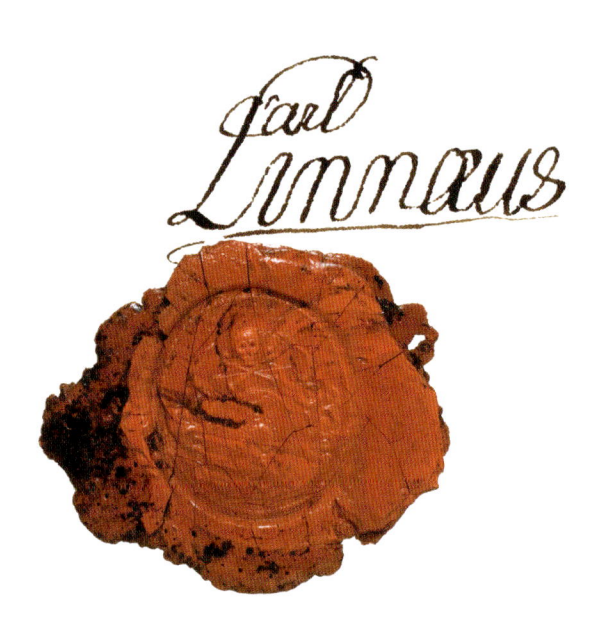

Dear Sir,

With some awe I dare, for the first time, approach you, one of the most renowned professors of all time at Uppsala University in Sweden. Even now, 250 years later, it is difficult to mention any researcher from here as globally famous as you have become. When I say 250 years later I count from the 1750s, the period when you had climbed to the top of your fame. Of course this was due to your extraordinary contributions in science, especially your numerous publications in botany. Your way of giving clues to the connections and relations between plants was adopted virtually everywhere and considered a great achievement to our knowledge of Nature. In addition you also discussed successfully the place of man in the zoological system. To the surprise of some of your colleagues, you were brave enough to put man in the hierarchy of animals and show the evidence for our background in the evolution of species. In the development of our understanding of the study of health and illness your contributions were also pioneering, alongside those in the science of minerals, what we now call geology.

Hammarby was Linnaeus' country house when he lived and taught in Uppsala

Part of your reputation in the world can be attributed to the travels of your students—the ones you liked to call your apostles. Like we are told from the Bible you sent your disciples around to 'preach' the new knowledge. Indeed they were very successful. With my own eyes I have witnessed that your student Carl Peter Thunberg has made an impact in Japan, just to mention one example. Your own name is found in many places, as well as monuments of different kinds. One example is in Amsterdam, where both a street and a square are named after you. In Barcelona, a corner of the Botanical garden keeps your memory alive by honouring you with a statue. In London, the Linnean Society takes extremely good care of some of your collections, those which were sold to a group of your admirers.

In Swedish schools, pupils were asked to collect flowers, identify them according to your system, dry and press them into a herbarium just like you did yourself. At least, this was the rule 60 years ago when I attended elementary school. I had to present a collection of close to 100 plants. The names and appearances of those plants always stayed in my memory and made it much more interesting to take walks in Nature. Your attitude towards what is found out there also stays: be careful and show respect for what you see! Today there is even a movement called ecology which is meant, among other things, to examine the impact of man in Nature.

My reason for writing this letter is to make you aware of a small contribution from your old university, intended to spread some information about you and your work to the whole world. It is not easy to explain how this information is distributed, but I shall do my best to describe the process.

A new technique has recently been invented whereby it is possible for me, sitting here in Uppsala, to communicate with the whole world. The only condition is that the people I want to approach also have the same equipment. However, the method is now so widespread that, at least in academic institutions, virtually everybody can be reached. The contact is immediate and therefore I can expect an answer in a very short amount of time, even from Japan, South Africa or a country in the American continents.

At Uppsala University we have written some texts which are accompanied by pictures, first in Swedish and then in English. English is now known almost everywhere—it has taken the place that Latin had in your time, to become the preferred language in scientific publications. Our texts are divided into seven sections, namely:

1) The Life of Linnaeus
2) Linnaeus and Pharmacy
3) Plants and Animals
4) Physics and the Cosmos
5) The History of Ideas
6) Linnaeus and Ecology
7) Mathematics in Linnaeus' Time

The authors are all well-known experts on the different topics represented. In this way we can proudly say that the information given is quite accurate and full of interesting details. My role in this project is to be the Editor which implies looking after all technical aspects, checking what is written and initiating extensions. At the start, over 10 years ago, I was already a Professor Emeritus at Uppsala University, with physics as my specialty. However, since people knew that I also had many other interests I was invited to become involved in 'Linné online', as the project is called. At that time there were five sections. In the meantime two more have been added. Another extension to the project was made rather soon after opening: we gave the readers out there the opportunity to contact the Editor and ask questions about the texts, or even to point out things which may be improved upon or corrected. The questions asked and the answers given are collected in an archive which is accessible to readers.

The number of readers is remarkable, between 500 and 1,000 per day throughout the year. In 2007, when we celebrated three centuries gone since your birth, the frequency was even higher. During the week around May 23rd our reader numbers increased to close to 2,000 per day. We know from where these readers come—not only Sweden; the whole world is represented.

CONTENTS OF THE SECTIONS

In the following I shall give a brief description of the contents of the seven sections of 'Linné online'.

In 1) we give your biography, from early childhood to your travels in Europe and in Sweden specifically, and your academic career in Uppsala. I should mention a curious thing about the date of your birthday. Due to a calendar reform at the end of the 18th century the date is no longer May 13th but May 23rd.

Later, I shall give a brief account of how your 300th birthday was celebrated. It has become clear that your biography contains many very important events, not only for your own personal part. From your books describing the travels in Sweden we learned a lot about our country and that even the smallest detail can play a role in our understanding, both of Nature and of our fellow beings. I should also mention that a book called *Vita Caroli Linnaei* has been published; it was initiated by Uppsala University and contains your five autobiographies and some very interesting explanations and translations of Latin texts.

Section 2) takes the reader to your best known specialty, plants and animals. It explains your system for characterizing the living world and also gives a description of some of the developments which have taken place since your time, mostly with a starting point in the results of your own research.

In section 3) your achievements in pharmacy are told as well as the status of pharmacy today. One can be surprised that even now, so long after your time, we take some of the basis for our drugs from the realm of plants.

Section 4) has the subtitle "What Linnaeus did not know about the Cosmos". Of course, in this particular research field much has happened in the last two centuries. Very little was known in your time about the smallest constituents of the natural world or about our universe, otherwise called the Cosmos.

Therefore, unlike the other sections, in this one it is almost impossible to make connections back to research in your century. Everything is new, and even in the years since the Linné online project started we have had to change certain things in the original text. That shows how fast this type of research is moving ahead.

Sections 5) and 6) have been produced by young university researchers in the discipline of history of science and ideas. Of course you take up a considerable amount of space in this field with your pioneering thoughts of the structure of the natural world. There are also many connections to our time.

Finally section 7) gives an account of the activities of one of your professorial colleagues at Uppsala University, and your friend, Samuel Klingenstierna. He was known all over Europe among mathematicians. In addition, he became the first professor of physics at our university. In this context I can mention that we have found a quotation from your writing about a visit by Klingenstierna at your country estate, Hammarby. It says that together you went for a walk in the surroundings, informally dressed. A well-known Swedish artist made

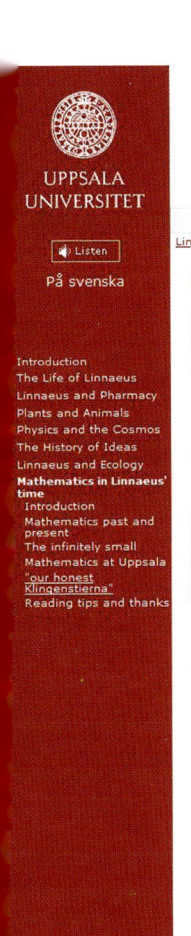

A page from 'Linné On Line' with a painting of the meeting between Carl Linnaeus and Samuel Klingenstierna

a colourful drawing of this moment and we have managed to include this in the section.

Well, dear Archiater and Professor at Uppsala University, I ask you to forgive me for my audacity in taking up your time for so long. However, it felt important to tell you that your life and work are still far from forgotten. Modern means have made it possible to spread the knowledge about your activities to an even larger audience than could be reached in your time.

Let me finish by telling you that your 300th anniversary in 2007 was celebrated in an impressive way. In Uppsala on the 23rd of May there were several manifestations: in the cathedral, in the university auditorium and in the castle. Among the many prominent guests were not only the Swedish King and Queen but also the Emperor and Empress of Japan, a really unique event!

Truly yours

Gunnar Tibell

Gunnar Tibell

Professor Emeritus at Uppsala University

Dear Professor Linnaeus:

I salute you as the developer of our system of naming organisms, now known as binomial nomenclature. It is such a simple concept, but you are to be congratulated for coming up with a two word name in either Latin or Greek, to designate each of the plants and animals. Without this system biological science would be in hopeless chaos.

There are so many plants and animals that it is difficult to coin a descriptive name for each, so we have resorted to using patronyms to honor individuals, or to indicate the origin of the organism. I must admit that I am flattered to have insects named for me, but my name on one can be misleading, leading one to think that it has three horns.

I will forever be indebted to you for coining the name "Tenebrio". It is the type genus of the beetle family Tenebrionidae, a group on which I have been working for more than half a century. I don't know how many genera and species you have described, but my mentor, Prof. Josef N. Knull, once told me that it is not how many taxa you describe, but how many withstand the test of time.

The amazing thing to me is how you were able to obtain so many specimens from all over the world. With antiquated postal services, no telephones, no computers, no e-mail, few publications, I marvel at what you accomplished.

You had an advantage on the rest of us systematists—most of your names are still valid, since there were no older names to contend with. You created no synonyms unless you happened to describe the same species twice. Nevertheless, you are indeed the Father of Binomial Nomenclature and your name will stand forever.

Sincerely yours,

Charles A. Triplehorn

Charles A. Triplehorn
Emeritus Professor of Entomology
The Ohio State University
Columbus, Ohio

A tray of darkling beetles (Tenebrionidae) from the Linnaean collections

News from Spain

Querido Carlos,

Congratulations! The changes that you introduced in the naming of living beings three hundred years ago have stood the test of time. Back then you sent your brightest student Pehr Loefling, your favorite 'apostle', to sample the flora of Spain. Prior to accepting that commission Loefling had been in charge of your son's education. He came to Spain in 1751, studying the flora of Madrid until 1754. Given the difficulties he encountered then he certainly did more than could have reasonably been expected, gathering a collection of more than 1,400 species, 562 from the vicinity of Madrid[1]. As you very well know, Loefling was then allowed to participate in the Spanish expedition to establish the limits of the Orinoco River. He spent two years in Venezuela collecting and studying plants but unfortunately died there from fever at the early age of 27. Often, as in Loefling's case, bad luck plagues one even beyond death: Loefling's Spanish herbarium, left in the hands of a local botanist, was destined to become lost forever. Not even a single clue as to what became of his material remains. The botanist left in charge of his herbarium neither studied it nor allowed anyone else to work with it. Such mishaps, more common than not among Natural History students, also occurred to one of your contemporaries, John Hunter. He, an insatiable student and collector, founded a Museum in London with more than 14,000 specimens. After his death, his notebooks, drafts and manuscripts fell into the hands of his brother-in-law and since have been lost. This peculiar psychological characteristic shared amongst these types of people, is very well portrayed by a traditional Spanish proverb: "*Son como el perro del hortelano, ni comen ni dejan comer*[2]."

In spite of the effect of such tragic losses to our knowledge of Nature, your system, well received in Spain, soon became key to the work of pioneer naturalists bent on increasing not only their knowledge of Iberian flora but also that of the Spanish 'colonies'. Ample evidence of the increasing acceptance of your system can be seen in the herbarium and archive of the Real Jardín Botánico, Madrid, which contains, for example, the impressive iconographic collection of Celestino Mutis, a significant contribution to our knowledge of the flora of Colombia.

By the end of the 18th century the results of the application of your system, well established in the practice of Spanish naturalists, began to appear in written documents. From 1799 until 1804, our Museum, known then as the Royal Spanish Cabinet, published 21 volumes of the *Anales de Historia Natural*, with contributions by Alexander von

Humboldt, the Spanish botanist José Antonio Cavanilles and others, including a translation of some text you had written to accompany Loefling's *Iter Hispanicum*[3]. Your system is used consistently throughout the five years covered by the annals. And, in a previous work, Juan Bautista Bru illustrated the holdings of the Royal Cabinet, using your system even for the monstrosities produced by nature. To give you a glimpse of that productivity I am including herein a plate with the description of one of the animals included in Bru's work.

Myrmecophaga tridactyla L. (Giant Anteater) illustrated in Juan Bautista Bru's *Colección de láminas que representan los animales y monstruos del Real Gabinete de Historia Natural de Madrid* (1784–1786), by courtesy of the Museo Nacional de Ciencias Naturales, Madrid

The second important matter I should like to point out in this letter has to do with the supposed opposition between your practice of classification and Charles Darwin's theory of evolution. May I say that a similar opposition occurred after the Spanish Civil War (1936–1939)? The war, anecdotally treated in Ernest Hemingway's *For Whom the Bell Tolls* counter-posed, in some sense, 'fix-its'—those who wanted to maintain the *status quo*—and 'evolutionists', those who hoped to advance toward a renewed society. This counterposition was reflected in the way Biology was taught after the war. The fix-its, who won the war, felt that you, Linnaeus, better represented their way of thinking. So a special version of your concepts and practice predominated in Spain, so that in the teaching of biological sciences avoidance of Darwin's theories or evolution lasted until well into the middle of the 20th century. This was not the only case in which such a 'straw man'—a deformed version of your views—was erected. At times, a straw man was used to support a fix-it's point of view on biological entities, while at other times evolutionists used the straw man to attack the supposed fixism in your thought.

Obviously, both uses neglect the fact that you were aware of the variability of organisms at least at the species level[4], even though you were not an evolutionist. And the attack was directed erroneously towards your supposedly limited view of living organisms. Nowhere in your work or in your practice is such a limit to be found.

Dogmatism pervades various groups of Spanish society today. A subtle anti-evolutionism exists there, a belief in divine intervention, at least at the origin of life and the transition from ape to man, two critical points in evolution. By contrast, in the last 20 years there has been a reaction against you and the people that continue your work, the pursuit of knowledge of earthly diversity of life. A recent re-editing of the aforementioned *Anales de Historia Natural*, is introduced by a study describing your work as ". . . la poco brillante tarea de describir géneros y especies en la más vulgar y banal línea impuesta por Linneo"[5]. The author, a historian of science, should have known that your work, far from being vulgar and banal, in fact has proven to be foundational.

While many scientists think that the discovering and describing of species is not a very lofty scientific enterprise, and label this area of study as 'light science', such a misconception leads to a faulty evaluation of the scientific impact of taxonomic work. As a result, the discovery of new taxa is not highly valued by the Spanish Science Committees that distribute honors and finance practising scientists. The main criteria used to evaluate scientific papers are 'citations' in other science papers (not content, but frequency of quotations by others). It is known that taxonomic works are cited by a small community of colleagues who know the group under study. This system of citation ranking considers the lack of citations of taxonomic work as a measure of lack of importance. Even such an eminent researcher as Peter Lawrence—winner of the Príncipe of Asturias prize in 2007, the Spanish Nobel prize—is surprised by the absurdity of the 'formulaic precision' with which Spanish researchers are evaluated by their own colleagues who are responsible for the ostracizing of taxonomists. What is even more contradictory is that while the discovery and description of taxa are not well appreciated, the analysis of the variation in spatial and temporal presence/absence of taxa is considered a 'hard science', which would be nothing at all without good taxonomy.

I shall not carry on with the long list of studies that argue over the scientific content laid out in classifications, and the enormous wealth of knowledge gained by adopting your system of naming. Even so I cannot help mentioning the not very well-known paper by Eric W. Holman, which I think shows paradigmatically the objectivity and rationality of your approach. The paper was published in 1985 in the *Journal of Classification*, a publication devoted to mathematical and theoretical aspects of classification. In this work Holman analyzed the structure of your classification of plants, animals, minerals and diseases. In Holman's own words: *"the differences between the classification of animals and plants and those of minerals and diseases reflects evolutionary properties of the material classified . . . Consequently, the observable effects of evolution are strong enough to be detected in classifications constructed before the acceptance of evolutionary theory. . ."*. This observation points toward the objective, rational and unbiased way in which you studied nature.

At present, the biological audience is not yet well-acquainted with the works of the scientific historian Mary Winsor, especially with her disagreements with Ernst Mayr's interpretation of your work as being essentialist/typologist, as was commonly viewed in the last part of the 20th century. The trend in thinking, following Winsor, is that Mayr's interpretation is erroneous, constructed through the use of a 'straw man' to comfortably discredit your legacy. Mayr's interpretation may be, in its turn, discredited within a few years.

Before ending this letter, I should point out a new avenue of thought that could bring new perspectives on, and a more rigorous understanding of, the work of taxonomists. I refer to the objectivity and testability of the discovery and description of taxa, which could be introduced with the question: Is taxonomy a hypothesis-driven discipline? As I said, there is a suspicion in a certain sector of the taxonomic community that taxonomy is not regarded as a true science, or at least is considered a 'half-science', whatever that may mean, by researchers in ecology, molecular biology and other disciplines. As you know, proposals and tests of hypotheses are at the nucleus of every scientific endeavour. Hypotheses have a clear status in science within the realm of statistics, a speciality of mathematics that has been enormously developed, especially during the 20th century. One of the branches of statistics that relies on frequencies—the repeated occurrence of items or facts—to calculate the possibility that they may occur again. However, the description of thousands of species, according to their very (and perhaps unique) nature has been, and is being rendered outside of this branch of statistical frequency methodology. Had such species, represented by unique (therefore non-statistically analyzable) specimens been ignored, we would have had to throw away the majority of what we know about life on earth that comes from fragments, pieces and in extreme cases, single specimens of past and present species.

The new twist, a Bayesian approach, comes from the ideas of the Reverend Thomas Bayes, a British mathematician who was your contemporary. Although this approach has not yet been fully developed I think that taxonomy may base its scientific structure on such an approach, rather than on the frequency statistical methodology. It can be said that taxonomists implicitly evaluate: ". . . the probability of a particular specimen belonging to a new species after having taking into account the previous known variation in the group." So, taxonomic hypotheses are Bayesian guesses of *a priori* probability and as such proper scientific hypotheses. In any case, even though many species have been described using the statistical frequency methodology, many more species descriptions have an implicit Bayesian foundation. As no-one can deny, there are good and bad hypotheses—well or poorly formulated, clear-cut or fuzzy. The quality of hypothesis does not depend on the methodology employed. So the recurrent use of this Bayesian approach should not lead to the unacceptable proposal of new species based on poorly evaluated material.

Taxonomical hypotheses are exercises in comparative biology. To do good comparative work, descriptions should be exhaustive and standardized. This, not usually the case, can most often be seen in a general revision of the group under study, the subject matter of monographs, and the place for the best taxonomic work. Paradoxically, as monographs do not appear in the list of the Science Citation Index, a compilation of citations of works in other publications, thus they are not very much appreciated by science evaluators, at least in Spain.

I think that a bright future is awaiting us, your disciples. You founded a solid discipline that has been contributing to the accumulation of a vast amount of knowledge, thousands and thousands of species since your time. There is no reason to suppose that

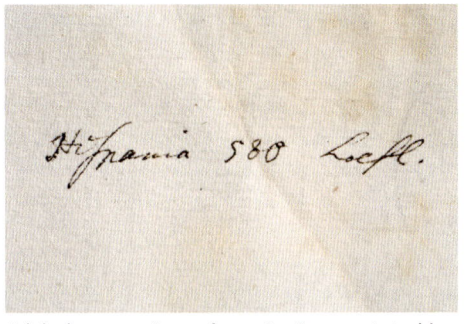

A label on specimen from Spain annotated by Linnaeus as being collected by Pehr Loefling

your programme will not continue to do so in years to come. As in Loefling's case, let us do as much as we can regardless of the unhelpfulness of colleagues.

Atentamente tuyo

Valdecasas

Antonio G. Valdecasas
Museo Nacional Ciencias Naturales
c/José Gutiérrez Abascal, 2
28006-Madrid. Spain.
Email: valdeca@mncn.csic.es

Notes:

[1] I learnt this about Loefling from my botanical colleague Ginés A. López Gónzalez, the person who knows best the early impact of your work in Spain.

[2] A kind of dog that 'neither eats nor allows others to do so.'

[3] You may check this in the section of Rare Holdings of the Inmaterial Taxonomic Library.

[4] See the thorough revision done by Müller-Wille in *Annals of Science* 64:171–215.

[5] '. . . the not very bright duty to describe genera and species in the most vulgar and trite style imposed by Linneo'. Joaquin Fernández Pérez, 1993. Los Anales de Historia Natural: entre un deseo real y una necesidad científica. Estudio preliminar, page 34. In: *Anales de Historia Natural*, Ed. Doce Calles.

Canterbury
Kent
England

22nd March 2008

Professor Carolus Linnaeus
Hammarby
SE-755 98 Uppsala
Sweden

Dear Carl

When I last wrote, some time ago, I pointed out that not only were your *Papilio* (*Barbarus*) *ancaeus* and *Papilio* (*Danaus*) *obrinus* male and female of the same species (just as Carl Alexander[1] had indicated), and then complained that you failed to spot this because your lousy characters led you to put them into different families—using families in Thomas's[2] sense (I still like to think that you intended your genera to be like suborders and the divisions to be families, and did not have in mind the solution your botanical pals came up with, making genera subordinate to families—but I digress). I went on to say that your system for the butterflies was just too crude, too superficial to get the real picture—as William[3] more or less said too.

I guess the fact you did not reply suggests you may have taken offence. Look, I was just getting a bit passionate, as usual, about my excitement over insect colour patterns and evolution (Henry[4] only managed to do such a convincing job with his mimicry idea because he took the time to look *behind* the shapes and colours, so to speak—while your work on the mammals was good enough to convince Johann[5] that we probably came from apes).

I know you are a proud man, and I really had no intention of causing offence. I should have been a bit less direct. So I am writing today to put the record straight, and reassure you that I really do appreciate what you have done for all of us—and by 'us', I mean the insects and the rest of nature too. Hell, we are all having a hard time with this population boom thing (more than 6 billion *Homo sapiens* and rising fast, far too fast in fact!), but if we didn't have the conviction that each kind of organism was more or less distinct, and deserved its own name, then we would have practically no idea what is at stake when it comes to this dreadful extinction business. At the time you were publishing '10' and '12' not even the idea of natural extinction in the geological past was contemplated, as John[6] has pointed out; now we have to deal with the shame that our "human enterprise" is causing thousands of extinctions every year—and rising as we get ever more out of sync with the rest of nature. But I digress again. What I really wanted to tell you was how I got into this whole business in the first place, and what your work really meant to me then—and still means to me now.

It all started on my seventh birthday, in July 1949. We were living in Orpington then (about four miles from Charles'[7] old place). An aunt of mine, Margaret Barrington, had given me a book on natural history as my birthday present. It was based around the idea of a country walk each month, right through the year, and the July walk was mostly about butterflies. With it was a coloured picture—six well-known sorts all feeding on *Buddleja*: Red Admiral, Peacock, Small Tortoiseshell, Large White, Small White, Meadow Brown. I went out into our back garden, where I remembered seeing such a bush—and there they all were. All the butterflies on the flowering *Buddleja* were in the coloured picture, and all the butterflies in the picture were on the bush! Well, so it seemed to me—and that was the important thing. I was hooked immediately: a clear link between nature 'outside' and knowledge 'inside'—inside a book in this case, but also in the author's head too, of course.

Soon after I started going to the public library, looking for more books on butterflies, and wandering over nearby 'waste ground', and in fields and woods, looking for the living creatures. I made a crude collection (pressed in an old book), and started raising caterpillars in jars. I found a Privet Hawkmoth caterpillar. It pupated, and I spent the whole winter and spring looking after it in a bowl of 'bulb fibre', longing to see it emerge. Eventually, one hot June day, I came home from school and found the empty chrysalis and the moth gone. I burst into tears, desperate at my loss. About an hour later I found the moth, perfect, clinging to the back of a curtain. There was joy in that moment!

The Privet Hawkmoth
(*Sphinx ligustri* L., Sphingidae)

A couple of years later (9th of August 1951 to be precise), I used pocket money I had saved up to buy *Butterflies and Moths of the Wayside and Woodland*[8]. My first 'serious' insect book—I still have it. From this I learnt that all species, in addition to their common English names, have two 'scientific' names, 'generic and specific'. The book also told me that the "number of known species of butterflies throughout the world has been put at about thirteen thousand, and it has been suggested that there may be nearly twice as many still awaiting discovery". That sounded pretty exciting. I was also introduced to 'the Palaearctic Region', which was said to have over 700 kinds, of which not more than 68 could "be regarded as British". What were the exotic species like? Amazingly, there was a retired priest living down the street, Rev. Dana, who had books on butterflies of India. I got to look at the coloured plates occasionally—so many wonderfully different-looking things, but some were familiar, or at least similar to our few species. That was exciting too.

But back to the names! Red Admiral: *Vanessa atalanta*. Peacock: *Nymphalis io*. Small Tortoiseshell: *Aglais urticae*, Large White: *Pieris brassicae*. Small White: *Pieris rapae*. Meadow Brown: *Maniola jurtina*. Privet Hawkmoth: *Sphinx ligustri*. Fascinating yet mysterious. Soon I was getting interested in other insects—beetles, bees, parasitic wasps, flies, grasshoppers, dragonflies. Then other pursuits started to compete: model aeroplanes, Meccano, making crude telescopes to wonder at the Pleiades, chemistry (nothing was more beautiful to me than a blue crystal of copper sulphate). Finally, there was music—especially Charlie and Dizzy[9]. But here I exceeded my capacity. It eventually dawned on me that I could never emulate the sheer creativity and vibrancy of their playing.

About the time of my 18th birthday, when I was finishing school, I hit on the idea of trying to work at the Natural History Museum in London. For years I had gone there to look at the butterfly collection in the public gallery, after finding a specimen of the Black-veined White (long extinct in England, I later realised it was accidentally released by a local butterfly breeder). On 4th April 1961 (with my *Wayside & Woodland* copy still less than 10 years old) I became an Assistant (Scientific) in the Museum's Department of Entomology. Starting out on Diptera, later switching to butterflies, and guided by wonderful colleagues like Norman, Paul, Harold and Ken[10], the mystery of the names was revealed. My six *Buddleja* butterflies, the Black-veined White, my great hawkmoth, you named them all—including the *Buddleja*! And even the two names thing, genus and species—that was your idea too. Amazing. Well, more or less ever since, I have been working away 'at the names', so to speak. Turns out for the butterflies, by the way, that there are an awful lot more than when you finished writing about them (you gave us names for 300[11]; currently we recognise about 20,000[12]; I think about 25,000 might be about the right final figure, if ever there is one). Even so, you should be proud of the fact that you described species belonging to all of the families currently recognised[13]—so, just as William[3] said, you really did lay the foundation for all that we know now.

'Reds': the Red Admiral, *Vanessa atalanta* (L, 1758) has a vast range across much of the northern hemisphere, but is divided into only two subspecies—one in America, the other in Eurasia. At one time the Red Admirals of Tuscany were thought to represent another race, but they do not. When the Indian Red Admiral, *V. indica* (Herbst, 1794), was first discovered in China around 1775, it was thought to be just a variety of *atalanta*. Nine species of 'reds' are now recognised worldwide[14]; the most recently discovered, from Timor, was only named in 1992. Top left: Canterbury, England; top right, Lexington, Ky, USA; lower left, Siena, Tuscany; lower right, *V. indica*, Kerala, southern India.

So, despite several false starts along the way, I have already spent nearly 60 years in the 'names business'—and, if good fortune continues to smile on me, I will continue for quite a few more years yet. Just a few weeks ago I published my first paper on the red admirals[14]—they still excite me, every time I see one (I've put a few of my own pictures of 'reds' in with this letter—why did you call our one *atalanta* by the way?). Although I have quite a few entomological heroes (Henry[4] and Willi[15] among them), we all owe so much to you. I will be your follower to the end of my days.

I must finish now. I have to prepare for a visit by my good friend Osamu Yata—he's coming over, all the way from Japan, to work with me and John Chainey. We are describing two new *Papilio* (*Danai candidi*)—well, *Appias* as we now refer to them. One is from a small group of islands in the Pacific, and hasn't been seen since the time I was born: we fear it may be extinct. But once we have given it a name—one of your good old binomens—people will really look out for it. Maybe it is still there and, if so, perhaps somebody will take care of it. Once species have names, it makes a big difference to how we feel about them. And we can communicate about them using the names too—what's outside is also inside our heads. *Vanessa atalanta*—that really means something to me, and many hundreds of other entomologists. *Vanessa atalanta* also means something in the environment—it has a real character, a real role in nature. For me, that seems to sum up the truly fantastic legacy you have given us all.

With all best wishes and, as ever, my deepest respects,

Sincerely,

R.I. Vane-Wright

PS. I hope your toothache is better now. I guess it must be. D.

Notes:

[1] Carl Alexander Clerck (1709–1765), whose contemporary illustrations are very important for understanding some of Linnaeus' insect species. On plate 31 of the second volume of his *Icones Insectorum Rariorum*, published in 1764, Carl Alexander illustrated both sexes of the South American butterfly *Nessaea obrinus* under the name *Papilio obrinus*.

[2] Thomas Pattinson Yeats (died 1782). Outstanding but little-known 18th-century English entomologist and follower of the Linnaean method. On page 132 of his *Institutions of Entomology*, published in London in 1773, Thomas wrote: "The division of the Butterflies [Linnaeus' *Papilio*] into families, from the circumstances chosen by Linnaeus, seems liable to many objections." He goes on to note that "Scopoli and Geoffroy have divided this genus into different families principally from the number of their feet" and, among other innovations, establishes to all intents and purposes the skippers based on their broad heads and unique antennae, and the swallowtails based on their larval osmeteria—and noting on this basis that the Apollo belonged with this group (using Linnaeus' name, the *Equites*), and not in *Heliconius*, where the great man had put it.

[3] William Jones of Chelsea (1745–1818). Jones was a wealthy London wine merchant, natural historian and scholar who 'retired to Chelsea', where he lived at No. 10 Manor Street. Even though it was never published, he is now mainly remembered for his '*Icones*', comprising some 1,500 watercolour images of butterflies and moths, now deposited in the University Museum, Oxford. Jones, however, did write one very important paper on butterfly classification (Jones, W. 1794. A new arrangement of Papilios, in a letter to the President. *Transactions of the Linnean Society of London* 2: 63–69, 1 pl.), in which he praised Linnaeus' foundational work, but also pointed out several shortcomings.

[4] Henry Walter Bates (1825–1892). Outstanding naturalist and entomologist who spent 11 years exploring the Amazon. On his return to England in 1860 he was able to interpret many of his discoveries in light of the new Darwin-Wallace theory of evolution by natural selection, and proposed his own special theory of false warning coloration, or *mimicry*. Bates' theory went a long way toward understanding many of the remarkable similarities in colour patterns between what we now know are only distantly related

species, a form of convergence that had caused much confusion to the early butterfly taxonomists.

⁵ Johann Christian Fabricius (1745–1808). Arguably Linnaeus' greatest student, Fabricius laid the foundations of modern insect taxonomy. He was, however, not just an entomologist, but a man of wide interests and achievements. His writings reveal that he believed new species and varieties could arise through hybridization, environmental influences and sexual preferences. In one passage, published in 1804, he even went so far as to suggest that *Homo sapiens* appears to have evolved from 'the bigger monkeys'—an idea that he undoubtedly got from studying the work of his great mentor.

⁶ Professor John Hedley Brooke. Former Director of the Ian Ramsey Centre, Oxford, and onetime editor of the *British Journal for the History of Science*. His outstanding *Science and Religion: Some Historical Perspectives* (Cambridge University Press, 1991) includes sections on Linnaeus and many of his contemporaries, and provides excellent insights into such topics as natural theology and the implications of extinction.

⁷ Down House, Downe, Kent, home of Charles Darwin.

⁸ *Butterflies and Moths of the Wayside and Woodland*, compiled by W.J. Stokoe from Richard South's three volumes on the butterflies and moths of the British Isles. My copy was the 1949 reprinting of the 1945 second edition, published by Frederick Warne, London.

⁹ Charlie Parker and Dizzy Gillespie—co-leaders of, arguably, the greatest quintet in the history of jazz. Parker played alto, Gillespie the trumpet. I could not afford a sax, which I wanted to play above all, so became a cornet and then trumpet player by default. They were heroes, and still are. Parker was already dead when I became aware of him, but I heard Dizzy live in London several times. Magic moments!

¹⁰ Norman Riley, Paul Freeman, Harold Oldroyd and Ken Smith. In their different ways, they taught me so much in my early days in the 'BMNH'. I probably learned more entomology and basic natural history from Ken than anyone before or since. He taught me a few other things too!

¹¹ Honey, M.R. & Scoble, M.J. 2001. Linnaeus's butterflies (Lepidoptera: Papilionoidea and Hesperioidea). *Zoological Journal of the Linnean Society* 132: 277–399.

¹² Kristensen, N.P., Scoble, M.J. & Karsholt, O. 2007. Lepidoptera phylogeny and systematics: the state of inventorying moth and butterfly diversity. *Zootaxa* (1668): 699–747.

¹³ Vane-Wright, R.I. 2007. Linnaeus' butterflies. *The Linnean Special Issue* 7: 59–74.

¹⁴ Vane-Wright, R.I. & Hughes, H.W.D. 2007. Did a member of the *Vanessa indica* complex formerly occur in North America? (Lepidoptera: Nymphalidae). *Journal of the Lepidopterists' Society* 61(4): 199–212.

¹⁵ Willi Hennig (1913–1976), founder of 'phylogenetic systematics', and progenitor of cladistics. I never met Willi Hennig, but he was the greatest influence on my early development as an insect systematist. The 'penny dropped' for me in 1966, when I was working in the BMNH during my long vacation on various Tipulidae related to and including *Holorusia*—the genus which includes the largest species of craneflies known. In 1963 I had stumbled across the fact that the generic name had been misapplied, and decided in 1965 that I would try to write this up, while I was still an undergraduate at University College London. On reading the 1966 English translation of Hennig's *Diptera Fauna of New Zealand as a Problem in Systematics and Zoogeography*, I was amazed to discover Hennig had concluded, based on his knowledge of zoogeography, that the current classification of these flies must be wrong, *even though he was totally unfamiliar with the insects concerned*. Such insight and conviction, in equal measure, made a profound impression on me, and I was soon wading through *Phylogenetic Systematics* (which appeared in translation the same year) to try to understand his methods.

16 June 1760

Viro Illustrissimo
Celeberrimo dom.
Dom. Carolo Linnæo
 Equiti &c
S.P.D.
 A. Vosmaer

Ex litteris Tuis perquam honorificis lætus intelligo meas ad Te pervenisse, sed miratus, quod in describenda Lacerta Anguina oblitus fuerim Aurium. En, illustrissime Vir! quæ de illis, aliisq insuper partibus notanda habeo.

Lacerta Anguina. Aures transversali et parum obliquo hiatu subangusto, angulis oris horizontali patentes.

Pedes omnino squamis subacutis obtecti sunt, unde pinniformes videntur.

Linea, quam in anteriore citavi, post accuratiorem intuitum nobis patet esse sulcus seu scissura longitudinalis singulis illis squamis insculptus.

Ad Lemurem Caudatum quod attinet, an non sit Mus volans, Syst. Nat. 63. N. 16, quæris? Certissimus respondeo, neutiquam: aliud plane est animal, et quantum scio a nullo auctore descriptum. Antequam Ego Musæo nostri Sereniss. Principis præfuerim, huic missum fuit, procul dubio ex Indiis orientalibus, sub nomine *Felis volantis*, cujus quoque animalis primo aspectu non absimile est; caput tamen magis acute protractum habet, ut ex addita hac delineatione apparet. Longitudo hujus animalis ab ore usque ad caudæ radicem est XV pollicum, mensura amstelod. cauda autem, quæ uti Felium rotunda est, XIX β vel XX pollicum est longa; illam in Delineatione nimis brevem exhibuerunt. — Ad pedes anteriores cutis parum reflexa hæret (etsi non tantum, quantum hic delineatum est) cutisque hæc in medio intra pedem anteriorem et posteriorem utriusq lateris ultra quinque pollices expanditur. Dubito an sat alte in ore unus vel alter dens molaris ad utrumq latus adsit, necne adesse sane videtur.

Argus Coluber, Seba Tom. 2. t. 103. f. 1. in Musæo nostro non invenitur, neq unquam a me visus.

Præcedentibus meis Litteris, circa Suecica sommodo mineralia et Lapidefacta supplex rogavi, ut si Tibi est occasio, illa Musæo nostro cura Tua acquiram, quum Suecia Regnum hic deficiunt adhucdum; lubenter ego transmissurus id quod a Te desideratum possidemus. An Clariss. D. Wallerius, indefessus ille Lapidei Regni scrutator, cujus Systema mihi valde acceptum est, in hoc negotio suam operam mihi præstare non poterit?

Nova Litteraria, quæ mecum communicare placuit, uti magni momenti, ita mihi grata fuisse gratissime agnosco; ast simul contineri doleo, Bibliopolas indies prodeuntia Tua opera huc non transmittere, ubi intra nova nil nisi Systema tuum prostat: avide jam expectamus tomum illum tertium, in quo non pauca quoq ut, in animali et vegetabili Regno, mutata et adjecta credimus. Speramus etiam illi additam fore accuratam Auctorum a Te laudatorum recensionem, uti et Nominum propriorum explicationem, id quod utilissimum futurum mihi videtur.

Dicas, quæso, an Belgice in posterum scribens, intelligi potero, citius sane honorifico tuo commercio tum satisfacere posse gaudebo.

In præsentiarum nihil est quod addam: ergo valeas, vir illustrissime, et me ex asse tuum ames.

Dabam Hagæ Com:
d. 16. Junii 1760.

Arnout Vosmaer (1720–1799) was a Dutch zoologist who was an early follower of Linnaean classification. Here he writes to "The Illustrious and celebrated Sir Carolo Linnaeo" about the identity of various specimens in the collection he was engaged in cataloguing (probably that of Adriaan Vroeg of The Hague) on the 16th of June 1760, shortly after the publication of the 10th edition of *Systema Naturae*.

Vosmaer describes a series of animals—"Lacerta anguina" etc., and suggests that various of the names applied by both Linnaeus and his contemporary Albertus Seba to flying mammals (*Lemurus caudatus*, *Mus volans*, *Felis volantis*) could be compared to good effect. He sends detailed descriptions of various of these, comparing them carefully with measurements and anatomical details, and includes a sketch for Linnaeus to examine.

Professor Carl von Linné
Most Honorable Swede and Citizen of the World

Dear Carl,

Please forgive my presumptively familiar salutation but I have admired and used your work for so long that I feel that I know you. Among my earliest memories in childhood was the use of Linnaean names for toy dinosaurs that I sought on 'expeditions' in miniature in my backyard. At an early age I discovered protozoans and spent much of my youth exploring pond and stream in search of species I had not yet seen, graduating to a copy of Richard Roksabro Kudo's *Protozoology* in ninth grade, and pure culturing ciliates by the gallon in a basement laboratory. It was the same year in my first biology class when I fully appreciated your achievements in classification and nomenclature and I have been a dedicated devotee ever since.

Placus (Prorodontidae) is a unicellular ciliate found in fresh and salt water

There are periodically those who, through ignorance or hubris or more often both, seek to replace the Linnaean system with another. One such proposal seems to be dying an appropriate but annoyingly slow death as I write. Called the PhyloCode, it represents the audacity of the first generation of systematic biologists arrogant enough to seek enshrinement of their ideas about relationships in a rigid system immutably imposed on future generations. Much greater minds, including yours, have pondered these issues and arrived at far better reasoned conclusions. What detractors fail to appreciate is that the persistence of the Linnaean system by practicing taxonomists is not an accident, not merely the lack of modernization or replacement. It is a calculated, deliberate perpetuation of a system that is so brilliant in its simplicity, logic and resilience that it is fully as useful today as it was when first penned by you in the 18th century.

Carl, I wish that you could see the sweeping changes that have taken place since your departure from this little-known planet. The theoretical landscape for taxonomy has shifted dramatically time

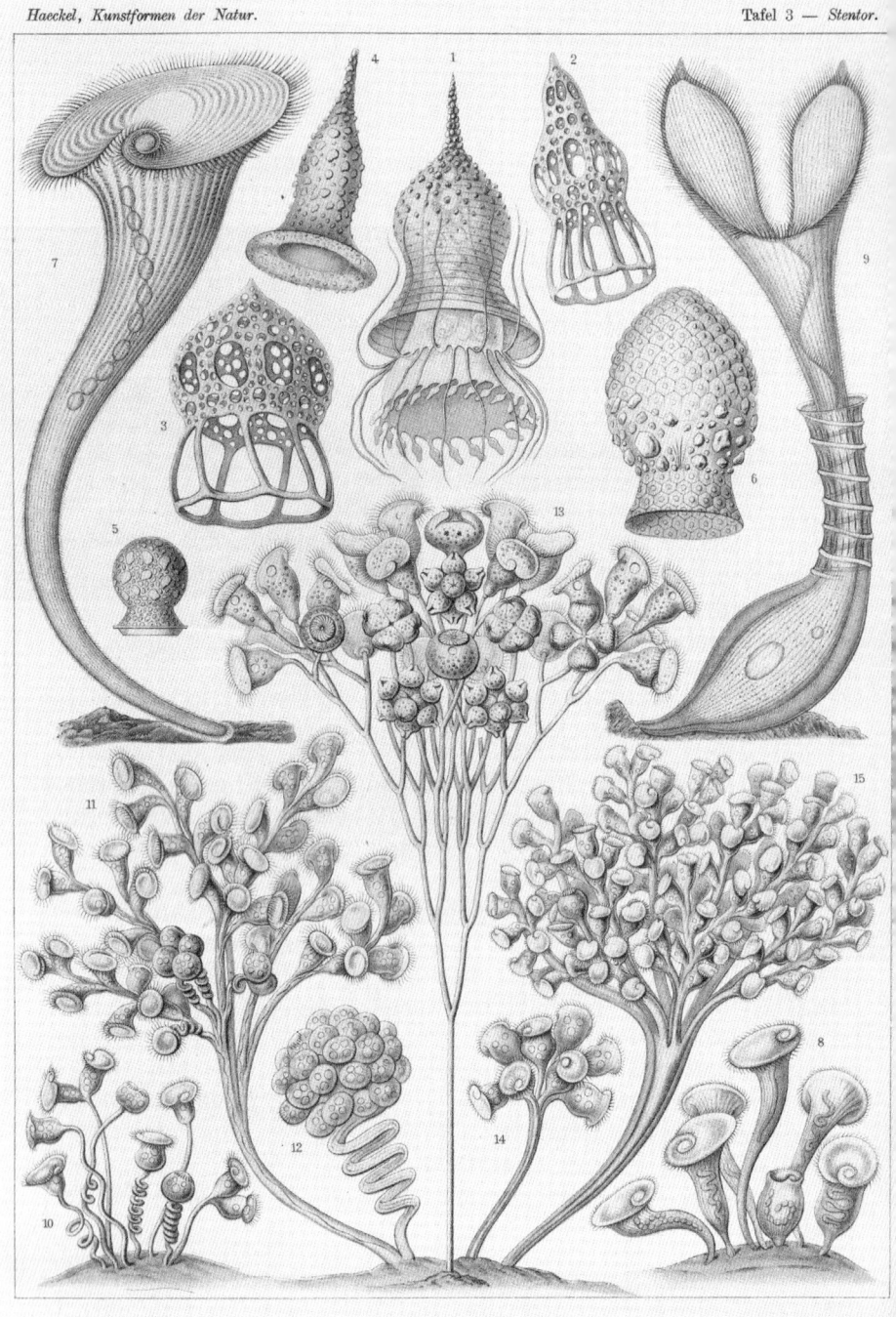

"Ciliata" (Table 3) in Ernst Haeckel's *Kunstformen der Natur* (1899)

and again—from the Creationist views that you shared with your contemporaries to Quinarians, Darwinians, Neo-darwinians, Evolutionary taxonomists, Pheneticists, Cladists, and now what might be termed Molecular Neo-pheneticists—and through it all it was the Linnaean system that gave clear and precise voice to each and every one of these world views. Your system is sometimes criticized for being imprecise in regard to reflecting phylogenetic relationships. This is a wrong-headed reading of the facts for at least two reasons. First, your system is flexible enough to reflect phylogenies pretty darned well even as our depth of understanding increases. An impressive amount of phylogenetic information can be conveyed through your system with the addition of minor devices, such as the phyletic sequence classifications advocated by Toby Schuh and others in the 1970s. Second, and more importantly, our knowledge of phylogeny is at best no more than approximately right so that a set of names that more or less convey phylogeny is all that we need. What is most important is that your system allows for constant adaptation as ideas change and new evidence is revealed. No one can deny that the hierarchic system you proposed was beautifully pre-adapted to be used when Willi Hennig's phylogenetic theories came along.

The other genius in your system, as if you did not know, was the introduction of binomials for species names. With millions of species, that simple device allows us to use memorable and descriptive adjectival words time and again in taxon after taxon. Those who question the wisdom of clinging to binomials have either failed to think this thing through or are not dealing with more than a handful of species.

What would sadden you, Carl, is to learn what is happening to the species on our planet that you so loved. Since your great works we have found that the Earth is home to millions of species more than the ten thousand or so known to you. We have observed that vast numbers of them are very limited in their distributions geographically or ecologically. We have recently come to appreciate just how rapidly natural ecosystems are being altered, degraded, and destroyed around the globe. As a result, millions of species face a very uncertain future and many of them are destined for almost certain extinction within the next few decades. Suddenly your species exploration enterprise, the original 'big science' project of the life sciences, has attained great urgency. While you were privileged to be the first generation to satisfactorily explore our planet's species, I have the misfortune of being a member of the last genera-

Orectochilus orbisonorum Miller, Mazzoldi & Wheeler (Gyrinidae): this whirligig beetle from India was named for musician Roy Orbison and his widow Barbara Orbison

tion with the option of discovering, describing and making known to mankind many of the unique species of our world.

You were a master not only of descriptive taxonomy, innovative nomenclature, and clever classifications, but also of inspiring students and gaining wide recognition and support for your work. I am hopeful that my generation can find within itself that same sense of curiosity and wonder and delight that accompanied the exploration of species in your time. By every measure the public seems open to such a message and I believe would be receptive. Regrettably, the obstacle to progress in species exploration and a Renaissance in Linnaean taxonomy involves colleagues within the biological sciences. Because experimental biology has come to dominate our universities and funding agencies, it is extremely difficult to receive funding or respect for descriptive, comparative, historical studies like taxonomy. This is a sad reflection on our education system that a student can receive a doctor of philosophy degree without enough introductory philosophy to even tell science from non-science. Just as your sexual system of classification for flowering plants was greeted by a grateful blushing public, it is increasingly clear that the necessary revival will come through a movement taken directly to the public and the user communities that depend so much on taxonomists for the ability to identify and communicate information about species. None of that is possible without sound, constantly tested classifications and species hypotheses, nor without Linnaean names.

There is great lamentation about the biodiversity crisis and a general sense of defeatism. This melancholy leads to complacency when in fact we need to be reminded of your ambitious works and thereby spurred to action. You did not shy away from the thousands of species that you had to deal with personally in your career but rather attacked the problem with great energy and leadership. With the advances made by you, Charles Darwin, and Willi Hennig we have the best theoretical foundation for classifications in history. With rapidly evolving digital technologies, we have the opportunity to adapt and invent our way around each and every impediment to rapid progress in taxonomy. We should be optimistic about what we can accomplish in regard to species exploration and thankful that we have the chance to document hundreds of thousands of species soon to be lost to science and the world. You had the prescience to begin this intellectual and scientific sojourn; we now need the vision to contribute to it in ways that no subsequent generation will be able to at any cost. No priority should be greater for society than exploring and learning the species with which we share this planet. Specimens and observations collected and taxonomy completed will arm us and future citizens of the world with the best hedge against uncertainty on a changing globe: knowledge. With a baseline of taxonomic knowledge we will be better able to conserve, manage and influence the state of biodiversity. And for the many species that will go extinct in spite of our best efforts, we shall create a permanent legacy of knowledge about the diversity of species on this little-known planet. This legacy, in the form of collections and taxonomic knowledge organized in Linnaean classifications, will continue to inform and inspire curious humans for countless generations to come.

Thank you for your system, your vision, and your example. Two hundred and fifty years after you published the 10th edition of *Systema Naturae* and ushered in the modern age of zoological nomenclature, you continue to inspire and to lead. I wanted you to know that you and your grand scheme are not forgotten and that many loyal disciples continue to walk in your footsteps and carry on your work. It has been an honor and privilege to have been such a foot soldier in the Linnaean army. To the extent that my generation succeeds in reinvesting in taxonomy in order to explore and learn our world's species, it will have been in no small part a direct result of your life and works. Thank you for making so many species available to us to observe and discover for ourselves and for your audacious vision to complete an inventory of our planet's species as a roadmap to meet our intellectual manifest destiny in understanding biological diversity.

With undying gratitude and admiration,

Your humble student,

Quentin D. Wheeler, FLS
University Vice President and Dean of the College of Liberal Arts and Sciences, Founding Director of the International Institute for Species Exploration, and Virginia M. Ullman Professor of Natural History and the Environment, Arizona State University

Dear Carl Linnaeus,

Gore Vidal (b. 1925), an American novelist, critic and social commentator, once wrote that "Each generation has its own likes and dislikes and ignorances"[1]. And so does each generation's scientists. Given the passage of time since publishing your various texts on classification, contrasting, as it were, the avowedly artificial 'Sexual System' for the identification of plants with the *Fragmenta Methodi naturalis*[2], your early attempt at a natural method of classification, these accounts are seen, with much justification, as placing "the study of systematic botany on a new and modern basis. The central problem, for him [Linnaeus] and for later workers as well, is the meaning and significance of artificial and natural systems"[3]. I thought it might be appropriate to consult you on these issues, given that today an odd, almost idiosyncratic, viewpoint has developed, offered for serious consideration, which suggests that because "Systematists get so worked up declaiming the centrality of classification in systematics . . . I [Joe Felsenstein] have argued the opposite"[4]—a position that has as a general philosophy "The irrelevancy of classification"[5]. Rather than worrying about various individual's likes and dislikes, is this, then, simply a case of ignorance? Let me first try and explain my own position, how I ended up needing to consult your various works.

I came upon your studies indirectly, not from the usual desire (or necessity) to find the name of a plant or animal, or even its source. I came through a desire to understand classification, the process, as it were, the intricacies of natural and artificial classifications, their relationship to one another and to the properties of the organisms we study. To achieve proper understanding of that subject, it was suggested I should begin by first reading *Fragmenta Methodi naturalis* and then Augustin Pyramus de Candolle's (1778–1841) *Théorie élémentaire de la Botanique*[6].

By the usual quirks of taxonomic endeavour I had already encountered the writings of A.-P. de Candolle. I study diatoms, single-celled photosynthetic organisms, the cell being encased in a silica shell. Their study requires a microscope. De Candolle had applied the generic name *Diatoma* to a few freshwater species in his *Flore Française* account (1805)[7], a genus I ended

Colliculoamphora reedii (Schrader) D.M. Williams & G. Reid (ca. 100 μm long); see Williams & Reid 2006. *European Journal of Phycology* 41: 147–154.

up studying in some detail[8]; de Candolle's name was eventually associated with the entire group, the Diatomaceae[9]. At the time of my studies on *Diatoma* I had no reason to consult either de Candolle's more general work, *Théorie élémentaire de la Botanique,* nor your *Fragmenta Methodi naturalis.*

Diatoms were apparently first recorded by a person identified only as a Mr C., a fellow of the Royal Society, who published his observations in the November–December 1703 issue of the *Philosophical Transactions*[10], some four years before your birth in response to a letter from the microscopist Antonie van Leeuwenhoek (1632–1723), who may have noticed their existence even earlier[11]. Leeuwenhoek noticed his particular diatom moving, so he assumed it to be an animal rather than a plant. In any case, he applied the term animalcule—'little animals'— a general term for all microscopic organisms. By the time you assembled information for the 12th edition of *Systema Naturae* (1767) these hybrid, 'composite' organisms were part of the Vermes, among the Zoophyta, a name conjuring up their composite nature (*zoon*, animal, *fyton*, plant). The Zoophyta turned out to be something of a mix, at first including such diverse organisms as *Volvox* and *Vorticella*, *Spongia* and *Hydra*, *Furia* and, in a manner of speaking as well as a taxon, *Chaos*[12]. Much later I happened upon a comment, attributed to you but that I have been unable to confirm, offering a prediction concerning this little piece of taxonomic chaos: "Mysterious living molecules, to be understood by our descendants"[13]. And so it was.

Diatoms were considered somewhat unusual organisms; some are motile *as well as* photosynthetic. They were considered animals by Christian Gottfried Ehrenberg (1795—1876), plants by Carl Adolph Agardh (1785–1859) and a combination of both by Friedrich Traugott Kützing (1807–1859)[14]. Ehrenberg, the "father of micropalaeontology"[15], treated them as 'polygastric' animalcules[16] ('rod' animalcules, *Bacillaria*[17]) among the infusoria—organisms so named as they seemingly arose *de novo* from infusions of hay. The name infusoria[18] was yet another collective term for the "mysterious living molecules". Still, diatoms were photosynthetic and could be treated as plants, placed among the algae, one of four orders—the others being mosses, ferns and fungi—in the 24th plant group of your sexual system, the Cryptogamia (from the Greek *kryptos*), those with reproductive organs hidden—"Nuptials are celebrated privately"[19]—a most decidedly artificial group.

Carl Adolph Agardh in his *Systema Algarum*[20] (1824) published one of the first comprehensive treatments of the algae. He classified diatoms in one order ('Ordo') of algae (Diatomeae), its species (for Agardh, around 47) distributed among nine genera. In the later *Conspectus Criti-*

cus Diatomacearum (1830–1832)[21], he re-arranged the genera into three families based on the shape of the siliceous valves: Styllarieae included genera with cuneate (wedged-shaped) valves, Cymbelleae included genera with cymbelloid valves and Fragilarieae included genera with rectangular valves. He compared each of these three families with four different colony 'types': those with no obvious colony formation ('libera'), those attached by a stalk ('Stipitata'), those attached in chains ('In frondem composita') and those in 'cymbelloid' chains ('Fila cymbellarum frondem formantia'), presenting his conclusions in tabular form, contrasting valve shape with colony structure. His classification, however, reflected the organism's frustule shape rather the formation of its colony. Agardh was interested in natural classification. He presented Augustin Saint-Hilaire (1779–1853) with a copy of his *Systema Algarum*: "I take the liberty to offer you a book on Algae. The idea that I have followed is not so much like those in present day systems that squeeze plants into a tidy frame, but rather like yours, to arrange them one nearer the other according to their greatest affinity"[22].

Agardh had written about affinity before in a little book called *Aphorismi Botanici*[23]. A year before Agardh published his *Systema Algarum*, the entomologist William Sharp MacLeay (1792–1865), in an essay on *The General Laws which have been lately observed to regulate the Natural Distribution of Insects and Fungi*[24], referred to Agardh's *Aphorismi Botanici* and *Systema Mycologicum*[25], published in 1821 by one of Agardh's students, Elias Magnus Fries (1794–1878), christened the "father of mycology" by Joseph Hooker[26]. MacLeay understood (and acknowledged) both Agardh and Fries as having independently developed ideas similar to his own, with respect to determining natural affinities; he wrote of their discussions of the pitfalls and problems in separating and distinguishing analogy from affinity, distinction being the key to finding the natural system: For Fries "Quo magis in superficiae acquieverunt naturae scruatores, co magis analoga cum affinibus commutarunt"; for Agardh "Analogia quaedam et similitudo in diversis seriebus vegetabilium interdum cernatur, quasi progressa esset natura ad perfectionem per cosdem gradus sed diversâ viâ".

Such terms—analogy and affinity—became much discussed, especially its association with discovering the 'Natural System' (natural classification). Agardh's viewpoint was of twofold interest, for diatomists at least. It marked the beginnings of the search for a natural classification of diatoms, and, somewhat ironically, almost its end, as all remaining efforts were tilted towards artificial classifications, 'systems' that might ease or assist identification, as if the only purpose for classifying diatoms was to enable general biologists to function more efficiently[27].

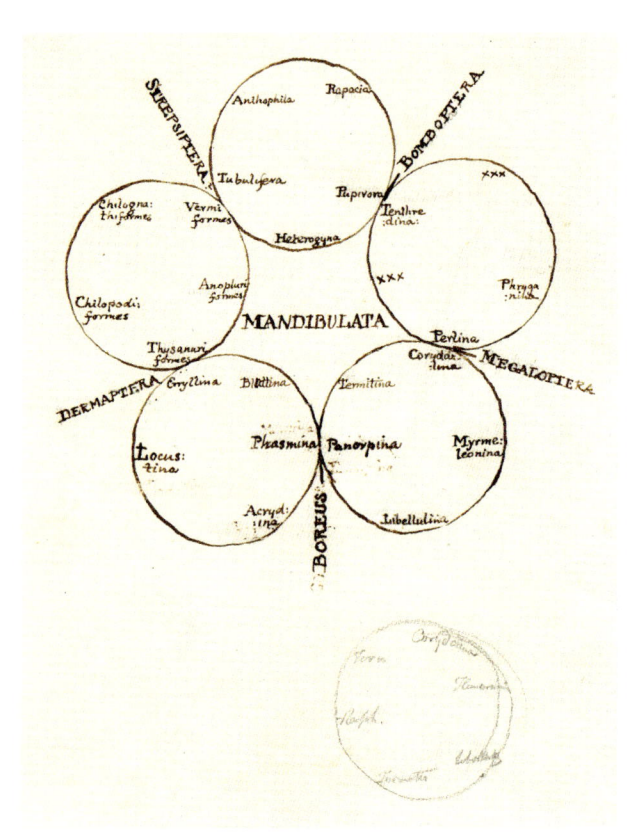

MacLeay gained a certain amount of notoriety, as he not only suggested the usefulness of separating affinity (homology) from analogy but offered the view that organisms were best placed in groups of five, a Quinary System of classification. Richard Owen (1804–1892), the first (and only) Superintendent of the Natural History Collections in the British Museum in London, wrote of MacLeay as having "the merit of first clearly defining and exemplifying, in regard to the similarities observable between different animals, the distinction between those that indicate 'affinity' and those that indicate 'analogy' or representation"[28]. Owen had earlier offered revised definitions of analogy and affinity, substituting the word homology for affinity, the term by which the relation is known today: homology is the key to the discovery of natural classification[29].

Manuscript drawing of MacLeay's Quinarian classification (see Williams & Ebach 2007. *The Foundations of Systematics and Biogeography*)

MacLeay mentioned a third person: Augustin Pyramus de Candolle and his monograph on the Cruciferae, published in *Memoire du Muséum d'Histoire Naturelle* in 1821[30]. MacLeay was interested in de Candolle's illustrations depicting the relationships, which could be expressed in a table showing "all the affinities existing in this family of plants by what he terms *double entrée*; in other words he supposes that there are transversal affinities as well as direct ones. . ."[31]. It was thus MacLeay and his discussion of classification that directed me to Candolle's *Théorie élémentaire de la Botanique*.

De Candolle's *Théorie élémentaire de la Botanique* began its life in the pages of the third edition of the *Flore Française* (1805)[32], which, many years later, Léon Croizat (1894–1982) described as "the first work in which the soul of the natural and artificial method had been laid bare"[33]. The earlier editions of the *Flore Française* were written entirely by Jean-Baptiste Lamarck (1744–1829); the first edition, in three volumes, was published in 1778; the second edition, in 1795, was almost identical to the first edition; but the third, with two additional volumes, was an entirely different work, from the pen of de Candolle alone. In the first volume of the third edition, de Candolle included an introductory text on the *Principes élémentaires de botanique*. He left Paris in 1807 for the University of Montpellier to become its professor of botany, where he began the book length treatment *Théorie élémentaire de la Botanique*; that work went into three editions, the first published in 1813, the second in 1819[34], and the third, published posthumously, in 1844[35]. An English translation was published in 1821, entitled *Elements of the philosophy of plants: containing the principles of scientific botany; nomenclature, theory of classification, phytography; anatomy, chemistry, physiology, geography, and diseases of plants: with a history of the science, and practical illustrations*, with Kurt Polycarp Joachim Sprengel (1766–1833) as co-author[36], derived from an earlier German edition, *Grundzüge der wissenschaftlichen Pflanzenkunde*[37] (1820). Of these translations, de Candolle noted that it was "M. Sprengel, qui unissant dans l'ouvrage ses idées aux miennes, et dans le titre son nom au mien, en fit un livre vraiment absurde où la fin de chaque chapitre est en opposition avec le commencement. . ."[38]. Thus, the English edition is not to be recommended—and nearly 200 years later we still await a good English translation. Nevertheless, in the *Théorie élémentaire de la Botanique*, de Candolle first referred to your own efforts: "Linnaeus was the first to distinguish carefully between the artificial method and the natural method . . . he was the first to give examples of one and the other."

In 1840 William Whewell (1794–1866) published the first edition of *The Philosophy of the Inductive Sciences*[39]. Within, book VIII was concerned with *Philosophy of the Classificatory Sciences*. Its chapter IV, *Of the Idea of Natural Affinity*, allowed Whewell to present a general principle of natural classification, a summary derived from de Candolle: "The basis of all Natural Systems of Classification is the Idea of Natural Affinity. The Principle which this Idea involves is this:—Natural arrangements, obtained from different sets of characters, must coincide with each other."[40] Natural classification, then, could be understood as a combination of homology and congruence: "The groups thus formed are related by Affinity [homology]; and in proportion as we find the evidence of more functions and more organs to the propriety of our groups, we are more and more satisfied that they are Natural Classes. It appears, then, that our Idea of Affinity [homology] involves the conviction of the *coincidence of natural arrangements formed on different func-*

Frontispiece from Richard Owen's *On the Nature of Limbs* (1849) designed to show his concept of special homology, the equivalence of parts

tions; and this, rather than the principle of the subordination of some characters to others, is the true ground of the natural method of Classification."[41]

But what of homology, its discovery and explanation? Owen rejected embryology (development), a popular method for determining homologues, and offered his archetype instead, an imaginary being from which all others are derived[42]. With the publication of *On the Origin of Species by means of Natural Selection* in 1859, Charles Robert Darwin (1809–1882) converted Owen's archetype into an ancestor, so that Owen's imaginary beast became a real, flesh and blood creature. Darwin spoke at length on the Natural System: "But many naturalists think that something more is meant by the Natural System; they believe that it reveals the plan of the Creator; but unless it be specified whether order in time or space, or what else is meant by the plan of the Creator, it seems to me that nothing is thus added to our knowledge. Such expressions as that famous one of Linnaeus, and which we often meet with in a more or less concealed form, that the characters do not make the genus, but that the genus gives the characters, seem to imply that something more is included in our classification, than mere resemblance. I believe that something more is included; and that propinquity of descent,—the only known cause of the similarity of organic beings,—is the bond, hidden as it is by various degrees of modification, which is partially revealed to us by our classifications."[43]

Darwin wrote on the subject of phylogenetics, the study of phylogeny, a neologism provided by the German morphologist Ernst Haeckel (1834–1919) to mean the "history of palaeontological development of organic beings"[44] (Haeckel also gave us 'ecology' and 'ontogeny'). Darwin added commentary to the fifth edition of his *Origin* (1869): "Professor Häckel in his 'Generelle Morphologie'[45] and in several other works, has recently brought his great knowledge and abilities to bear on what he calls phylogeny, or the lines of descent of all organic beings . . . He has thus boldly made a great beginning, and shows us how classification will in the future be treated."[46]

Darwin, inadvertently, spawned a wealth of -isms in the 19th century and, later, in the 20th century, spawned his own industry[47]. Here's one

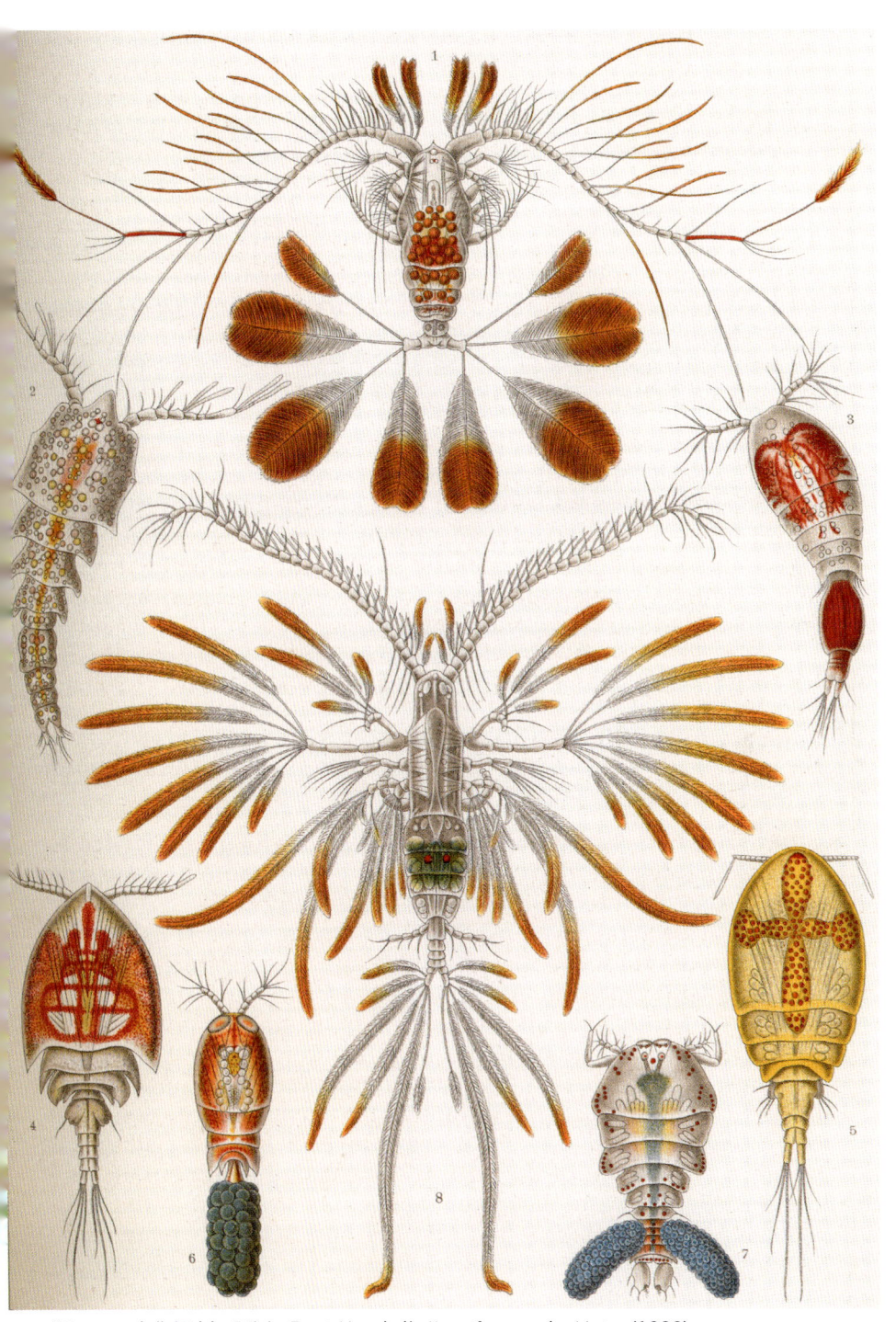

"Copepoda" (Table 56) in Ernst Haeckel's *Kunstformen der Natur* (1899)

contemporary comment from an introduction to a book entitled *Darwinian Heresies*: "The intellectual landscape of Darwinism for the last 150 years bears a certain resemblance to Germany during the Thirty Years' War. Sects and churches, preachers and dissenters, holy warriors and theocrats vie with each other for the hearts of the faithful and the minds of the unconverted, all too often leaving scorched earth behind. Such an extravagant metaphor is not much of an overstatement."[48] It's of little consequence what we may today believe Darwinism to be, or what it might have evolved into, for whatever form, in the latter part of the 20th century (and on into the 21st century) it has been discussed with reference to classification[49]—but like the palaeontologists of the 20th century, "Many phylogeneticists now see nomenclature and classification as largely irrelevant to phylogenetics. . ."[50]—a position that can only be arrived at out of ignorance.

Nevertheless, Haeckel's boldness did not go unchallenged, of course, with a sophisticated critique mounted mainly in Europe, beginning with Adolf Naef (1883–1949)[51], but for decades falling on the deaf ears of the Anglo-American development, which became somewhat preoccupied with fossils, searching directly, so they believed, for the "history of palaeontological development of organic beings". Meaningful discussion of classification was little in evidence.

Cladistics might be seen as a reaction to phylogeny, or at least a reaction to Haeckel's version of it[52]. That reaction appeared to have its birth in entomology[53] but its origin might really be with Naef[54], and the development of cladistics, somewhat oddly, within the discipline of palaeontology[55].

As it turned out, cladistics was quickly misunderstood, and viewed as if it was simply a method to discover phylogeny without necessarily using fossils (via the relationship of common ancestry rather than ancestor-descendant relationships)[56]. Yet among the first full-length treatments of cladistics[57], Gareth Nelson and Norm Platnick's *Systematics and Biogeography* differs, as it is a detailed critique of Haeckel's legacy, along with a cogent reassessment of natural and artificial classification, perhaps the first original effort since de Candolle's *Théorie élémentaire de la Botanique*[58]. In short, cladistics was a confirmation that natural classification requires homology and congruence; characters coinciding with each other. As summed up elsewhere: "If cladistics is merely a restatement of the principles of natural classification, why has cladistics been the subject of argument? I suspect that the argument is largely misplaced, and that the misplacement stems, as de Candolle suggests, from the confounding goals of artificial and natural systems."[59]

In truth, the cladistic revolution should not have been necessary. It should have been obvious, as cladistics was quite simply further commentary on Natural Classification—albeit long overdue. It should have returned our collective thoughts and efforts to the job of classifying the organisms in the world—the process set in motion, in its modern form, by *Fragmenta Methodi naturalis*. For the moment, that project has stalled a little.

As for diatoms, at a rough guess there are now some 10–12,000 species known—in contrast to the 47 Agardh had to deal with—which only really scratches the surface. And yet a natural classification still seems a long way off[60] but not because data aren't available or relevant or even in abundance. It's simply because artificial and natural classifications have never been recognised for what they are, and separated and used accordingly. When that happens, diatom systematics will fall into place, regardless of any new sources of data, regardless of either its quantity or quality[27].

Sincerely,

David M. Williams

Notes:

[1] Vidal, G. 1987. The bookchat of Henry James. In: *Armageddon? Essays 1983–1987*, Andre Deutsch, p. 167.

[2] Linnaeus, C. 1738. *Fragmenta Methodi naturalis*, in *Classes plantarum seu Systemata plantarum. Omnia a fructificatione desumta, quorum XVI universalia & XIII partialia, compendiose proposita secundum classes, ordines et nomina generica cum clave cujusvis methodi et synonymis genericis. Fundamentorum botanicorum pars II*, Leiden. See Larson, J.L. 1967. Linnaeus and the natural method. *Isis* 58: 304–320; Svenson, H.K. 1945. On the descriptive method of Linnaeus. *Rhodora* 47: 273–302, 363–388.

[3] Nelson, G. & Platnick, N.I. 1981. *Systematics and biogeography: cladistics and vicariance*. Columbia University Press, New York, p. 88.

[4] Franz, N. 2005. On the lack of good scientific reasons for the growing phylogeny/classification gap. *Cladistics* 21: 495–500, p. 495.

[5] Felsenstein, J. 2004. *Inferring Phylogenies*. Sinauer Associates, Sunderland, MA, p. 145.

[6] Candolle, A.-P. de 1813. *Théorie élémentaire de la botanique ou exposition des principes de la classification naturelle et de l'art de décrire et d'étudier les végétaux*. Deterville, Paris.

[7] Candolle, A.-P. de in Lamarck, J.-B. P. A. de M. & Candolle, A.-P. de 1805. *Flore française, ou, Descriptions succinctes de toutes les plantes qui croissent naturellement en France disposés selon une nouvelle méthode d'analyse, et précédés par un exposé des principes élémentaires de la botanique*. Volume 2, p. 48. H. Agasse, Paris.

[8] Williams, D.M. 1985. Morphology, taxonomy and inter-relationships of the ribbed araphid diatoms from the genera *Diatoma* and *Meridion* (Diatomaceae: Bacillariophyta).

Bibliotheca Diatomologica 8: 1–228; Williams, D.M. 1990. Cladistic analysis of some freshwater araphid diatoms with particular reference to *Diatoma* and *Meridion*. *Plant Systematics & Evolution* 171: 89–97.

[9] Dumortier, B. C. J. 1822. *Commentationes botanicæ: observations botaniques, dédiées à la société d'horticulture de Tournay*. Tournay: C. Casterman-Dieu.

[10] Mr. C. 1703. Two letters from a gentleman in the country, relating to Mr Leuwenhoeck's Letter in Transaction, No. 283. Communicated by Mr C. *Philosophical Transactions* 23(288): 1494–1501.

[11] Leeuwenhoek, A. 1703. Part of a Letter from Anthony von Leeuwenhoek, Concerning green weeds growing in water, and some animicula found about them. *Philosophical Transactions* 23(283): 1304–1311; Leeuwenhoek, A. 1996. Letter nr. 239 of 25 December 1702. In: *The Collected Letters of Antoni Van Leeuwenhoek* (L. C. Palm, ed.), volume 14: 158–179, Swets & Zeitlinger, Lisse; Ford, B. J. 1981. Leeuwenhoek's specimens discovered after 307 years. *Nature* 292: 407. Leeuwenhoek's letter is dated 25th December 1702 but was not printed in the *Philosophical Transactions* until early 1703.

[12] Linnaeus, C. 1767. *Systema Naturae, per regna tria naturae, secundum classes, ordines, genera, species cum characteribus, differentiis, synonymis, locis*. Vindobonae: Typis Joannis Thomae . . . Trattnern, p. 1287–1327; Jarvis, C. 2007. *Order out of Chaos: Linnaean Plant Names and their Types*. The Linnean Society of London, London.

[13] Quotation, which concerned ciliates, is found in Fokin, S.I. 2004. A brief history of ciliate studies (late XVII—the first third of the XX century). *Protistology* 3: 283–297, who gives as its source Sobol S.L. 1949. History of microscope and microscopical studies in Russia at XVIII century. Akademie Science USSR, M.L. (in Russian).

[14] Kützing, F.T. 1843. *Phycologia Generalis oder Anatomie, Physiologie und Systemkunde der Tange*. Brockhaus, Leipzig; Kützing, F.T. 1844. *Über die Verwandlung der Infusorien in niedereAlgenformen*. Köhne, Nordhausen.

[15] Williams, D.M. & Huxley, R. (eds). 1998. Christian Gottfried Ehrenberg (1795–1876): The man and his legacy. *Linnean Society Special Publications* no. 1; an accolade that might encourage dispute, as many have suggested Alcide d'Orbigny (1802–1857), the French biologist, has a more appropriate claim to that title: see Le Calvez, Y. 1974. Greatest names in micropaleontology. 1. Alcide d'Orbigny, p. 261–264, in Hedley, R.H. & Adams, C.G. (Eds.), *Foraminifera, Volume 1*. Academic Press, London; Vénec-Peyré, M.-T. 2004. Beyond frontiers and time: the scientific and cultural heritage of Alcide d'Orbigny (1802–1857). *Marine Micropaleontology* 50: 149–159 and the collection of papers in commemorative issues of *Compte Rendu Palevol*. 2002, 1(6) and 1(7).

[16] "Ehrenberg divided the Infusoria into polygastrica and rotatoria. The characteristic of the former is the appearance of certain internal cavities, which he suppose to be dilated portions of the alimentary canal, hence their name polygastic, or many-stomached." From Ripley, G. & Dana, C.A. (1858). *The New American Cyclopedia: A Popular Dictionary of General Knowledge* Volume 1, A–Araguay. New York, D. Appleton & Co., p. 604; the many 'stomachs' were actually vacuoles, see Taylor, F.J.R. 2003. The collapse of the two-kingdom system, the rise of protistology and the founding of the International Society for Evolutionary Protistology (ISEP). *International Journal of Systematic & Evolutionary Microbiology* 53: 1707–1714, p. 1708 and Churchill, F.B. 1989. The guts of the matter. Infusoria from Ehrenberg to Bütschli: 1838–1876. *Journal of the History of Biology* 22: 189–213.

[17] *Bacillaria*, as a genus of diatoms, was first included in the 13th edition of *Systema Naturae*. Linnaeus, C. 1788. *Systemae naturae per regna tria naturae secundum classes, ordines, genera species cum characteribus, differentiis, synonymis, locis* (13 ed.), Tomus 1, pars 6, p. 3903. Lemaire, Brussels; see Ussing, A.P., Gordon, R., Ector, L., Buczko, K., Desnitskiy, A.G. & VanLandingham, S.L. 2005. The Colonial Diatom "*Bacillaria*

paradoxa": Chaotic Gliding Motility, Lindenmeyer Model of Colonial Morphogenesis, and Bibliography, with Translation of O. F. Müller (1783) "About a peculiar being in the beach water". *Diatom Monographs* 5; Schmid, A.-M. 2007. The "paradox" diatom *Bacillaria paxillifer* (Bacillariophyta) revisited. *Journal of Phycology* 43: 139–15; Jahn, R. & Schmid, A. M. 2007. Revision of the brackish-freshwater genus *Bacillaria* Gmelin (Bacillariophyta) with description of a new variety and two new species. *European Journal of Phycology* 42: 295–312.

[18] Wrisberg, H.A. 1765. *Observationum de Animalculis Infusoriis Satura.* Vandenhoeck, Gottingen (Wrisberg's book probably contains the first usage of Infusoria; much useful discussion was supplied by Barry Leadbeater, pers. comm. on the use of Infusoria).

[19] Linnaeus, C. 1735. *Systema Naturae, sive, regna tria naturae systematice proposita per classes, ordines, genera & species.* Lugduni Batavorum: Apud Theodorum Haak; ex typographia Joannis Wilhelmi de Groot.

[20] Agardh, C.A. 1823–1828. *Systema Algarum.* Litteris Berlingianis, Lundae; the diatom section was published in 1824; see Ott, F.D. 1991. Carl Adolph Agardh, Professor, Bishop. A translation of J.E. Areschoug's 1870 Memorial. *Archiv für Protistenkunde* 139: 297–312.

[21] Agardh, C.A. 1830–1832. *Conspectus Criticus Diatomacearum.* Litteris Berlingianis, Lundae.

[22] Woelkerling, W.J. and Lamy, D. 1999. *Non-geniculate coralline red algae and the Paris Muséum: Systematics and scientific history.* Paris Publications Scientifiques du Muséum, A.D.A.C., Paris, p. 78.

[23] Agardh, C.A. 1819 [1817–1826]. *Aphorismi Botanici.* Literis Berlingianis, Lund.

[24] MacLeay, W.S. 1823. The General Laws which have been lately observed to regulate the Natural Distribution of Insects and Fungi. *Philosophical Magazine* 62: 192–200, 255–262; reprinted in *Transactions of the Linnean Society* 14: 46–68, 1825.

[25] Fries, E.M. 1821–32. *Systema Mycologicum: Sistens Fungorum ordines, genera et species huc usque cognitas, etc.* 3 volumes. Lundæ & Gryphiswaldœ. Volumes 2 and 3 are in 2 parts, pagination is continuous. Volume 2 is dated 1822–23 and volume 3 is dated 1829–32. Volume 1 includes 'De Systematis Constructione' in the introduction, Fries's discussion of natural classification.

[26] Hooker, J. D. 1876. *Address of Joseph Dalton Hooker, C.B., the President, delivered at the anniversary meeting of the Royal Society, on Thursday, November 30, 1876.* London: Royal Society; Korf, R.P. 2005. Reinventing taxonomy: a curmudgeon's view of 250 years of fungal taxonomy, the crisis in biodiversity, and the pitfalls of the phylogenetic age. *Mycotaxon* 93: 407–415.

[27] Williams, D.M. & Kociolek, J.P. 2007. Pursuit of a natural classification of diatoms: History, monophyly and the rejection of paraphyletic taxa. *European Journal of Phycology* 42: 313–319; Williams, D.M. 2007. Classification and diatom systematics: The past, the present and the future. In: *Unravelling the Algae* (eds. Brodie, J. & Lewis, J.), pp. 57–91. CRC Press.

[28] Owen, R. 1859b. Address. *Report of the 28th Meeting of the British Association for the Advancement of Science.* John Murray, London, pp. xlix–cx, p. lxvii.

[29] Owen, R. 1843. *Lectures on the comparative anatomy and physiology of the invertebrate animals, delivered at the Royal College of Surgeons in 1843, from notes taken by W. W. Cooper . . . and revised by Prof. Owen, etc.* London: Longman, Brown, Green and Longman.

[30] Candolle, A.-P. de 1821. Mémoire sur la famile des Crucifères. *Memoire du Muséum d'Histoire Naturelle* 7: 169–252; see Stevens, P.F. 1984. Haüy and A.-P. Candolle: Crystallography, botanical systematics, and comparative morphology, 1780–1840. *Journal of the History of Biology* 17: 49–82, p. 65.

[31] Macleay, W.S. 1823. The General Laws which have been lately observed to regu-

late the Natural Distribution of Insects and Fungi. *Philosophical Magazine* 62: 192–200, 255–262, quotation on p. 195, referring to de Candolle, A.-P. 1821, see footnote 30.

[32] Lamarck, J.-B. P. A. de M. & Candolle, A.-P. de 1805–1815. *Flore française, ou, Descriptions succinctes de toutes les plantes qui croissent naturellement en France disposés selon une nouvelle méthode d'analyse, et précédés par un exposé des principes élémentaires de la botanique.* H. Agasse, Paris.

[33] Croizat, L. 1945. History and Nomenclature of the Higher Units of Classification. *Bulletin of the Torrey Botanical Club* 72: 52–75, p. 64.

[34] Candolle, A.-P. de 1819. *Théorie élémentaire de la botanique . . .* 2nd edn. Deterville, Paris.

[35] Candolle, A.-P. de 1844. *Théorie élémentaire de la botanique . . .* 3rd edn. Roret, Paris.

[36] Candolle, A.-P. de & Sprengel, K.P.J. 1821. *Elements of the philosophy of plants: containing the principles of scientific botany; nomenclature, theory of classification, phytography; anatomy, chemistry, physiology, geography, and diseases of plants: with a history of the science, and practical illustrations.* Edinburgh; London.

[37] Candolle, A.-P. de & Sprengel, K.P.J. 1820. *Grundzüge der wissenschaftlichen Pflanzenkunde.* Leipzig.

[38] Candolle A.-P. de 1862. *Mémoires et souvenirs, écrits par lui même et publiés par son fils.* Joël Cherbuliez, Genève & Paris. Candolle A.-P. de 2004. *Mémoires et souvenirs (1778–1841).* Candaux, J.-D. & Drouin, J.-M. (eds.), Bungener, P. et Sigrist, R. (coll. eds.), Georg, Genève. (*Bibliothèque d'Histoire des Sciences*, vol. 5); Bungener, P. 2004. Mémoires et souvenirs d'Augustin-Pyramus de Candolle ou l'autobiographie d'un savant botaniste. *Archives de Science* 57:41–46.

[39] Whewell, W. 1840. *The Philosophy of the Inductive Sciences, Founded upon their History.* London: J.W. Parker. The second edition was published in 1847, the third edition, in three separate parts, from 1858–1860, with *Novum Organon Renovatum*, Part II of the 3rd volume of the third edition.

[40] Whewell, W. 1847. *The Philosophy of the Inductive Sciences, Founded upon their History.* London: J.W. Parker, p. 537.

[41] Whewell, W. 1847. *The Philosophy of the Inductive Sciences, Founded upon their History.* London: J.W. Parker, p. 542.

[42] Owen, R. 1846. Observation on Mr. Strickland's article on the structural relations of organized beings. *Philosophical Magazine* 28: 525 7; Owen, R. 1847a. Report on the archetype and homologies of the vertebrate skeleton, *Report of the 16th Meeting of the British Association for the Advancement of Science*, pp. 169–340. London: Murray; Owen, R. 1848. *On the Archetype and Homologies of the Vertebrate Skeleton*, London; Owen, R. 1849. *On the Nature of Limbs*, London: John van Voorst; Owen, R. 1866. *On the Anatomy of Vertebrates, Volume 1, Fishes and reptiles*, London: Longmans, Green and Co.

[43] Darwin, C. R. 1859. *On the Origin of Species by means of Natural Selection, or the preservation of favoured races in the struggle for life.* London: John Murray, pp. 413–414.

[44] Haeckel, E. 1874. *Anthropogenie oder Entwickelungsgeschichte des Menschen: Gemeinverstandliche wissenschaftliche Vortrage uber die Grundzuge der menschlichen Keimes- und Stammes-genchichte.* Wilhelm Engelmann, Leipzig, pp. 18 and 710.

[45] Haeckel, E. 1866. *Generelle Morphologie der Organismen.* Berlin: Georg Reimer.

[46] Darwin, C. R. 1869. *On the Origin of Species by means of Natural Selection, or the preservation of favoured races in the struggle for life.* London: John Murray. 5th edition, p. 515.

[47] Flannery, M.C. 2006. The Darwin Industry. *The American Biology Teacher* 68: 163–166; Ruse, M. 1996. The Darwin Industry: A Guide. *Victorian Studies* 39: 217–235;

Lenoir, T. 1987. Essay Review: The Darwin Industry. *Journal of the History of Biology* 20: 115–130.

[48] Lustig, A. Introduction. Biologist on crusade, in Lustig, A., Richards, R.J. and Ruse, M. (Eds), *Darwinian Heresies*. Cambridge University Press, pp. 1–13, quotation on p. 1.

[49] Mayr, E. 1974. Cladistic analysis or cladistic classification? *Zeitschrift für Systematik und Evolutionsforschung* 12: 94–128; Nelson, G.J. 1974. Darwin-Hennig classification: A reply to Ernst Mayr. *Systematic Zoology* 23: 452–458; Mayr, E. and Bock, W.J. 2002. Classifications and other ordering systems. *Zeitschrift für Zoologische Systematik und Evolutionsforschung* 40: 1–25; De Queiroz, K. 1988. Systematics and the Darwinian revolution. *Philosophy of Science* 55: 238–259.

[50] Hillis, D. M. 2007. Constraints in naming parts of the Tree of Life. *Molecular Phylogenetics and Evolution* 42: 331–338, quotation on p. 331.

[51] Boletzky, S. von 1999. Systematische Morphologie und Phylogenetik-zur Bedeutung des Werkes von Adolf Naef (1883–1949). *Vierteljahrsschrift der Naturforschenden Gesellschaft in Zürich* 144: 73–82; Boletzky, S. von 2000. Adolf Naef (1883–1949). A biographical note. In *Fauna and Flora of the Bay of Naples [Fauna und Flora des Golfes von Naepel]*. Monograph 35. Cephalopoda. Embryology. Part I, Vol. II [Final part of the Monograph No. 35], pp. ix–xiii. Smithsonian Institution Libraries, Washington, DC.

[52] Hennig, W. 1950. *Grundzüge einer Theorie der phylogenetischen Systematik*. Deutsche Zentralverlag, Berlin [reprinted 1980 Otto Koeltz, Koenigstein]; Hennig, W. 1957. Systematik und Phylogenese. Pp. 50–71 in *Bericht über die Hundertjahrfeier der Deutschen Entomologischen Gesellschaft, Berlin. 30 September bis 5 Oktober 1956*, (ed. by H. von Hannemann), Akademie-Verlag, Berlin; Hennig, W. 1965. Phylogenetic systematics. *Annual Review of Entomology* 10: 97–116; Hennig, W. 1966. *Phylogenetic Systematics*. University of Illinois Press, Urbana [Reprinted 1979, 1999].

[53] Hennig, W. 1953. Kritische Bermerkungen zum phylogenetischen System der Insekten. *Beiträge zur Entomologie* 3:1–85.

[54] Naef, A. 1917. Die individuelle Entwicklung organischer Formen als Urkunde ihrer Stammesgeschichte: (Kritische Betrachtungen über das sogenannte "biogenetische Grundgesetz"). Verlag von Gustav Fischer, Jena; Naef, A. 1919. Idealistische Morphologie und Phylogenetik (zur Methodik der systematischen); Verlag von Gustav Fischer, Jena.; see Williams, D.M. & Ebach, M.C. 2007. Foundations of Systematics and Biogeography. Springer-Verlag New York Inc.

[55] Nelson, G. J. 1969. Gill arches and phylogeny of fishes, with notes on the classification of vertebrates. *Bulletin of the American Museum of Natural History* 141: 475–552; Patterson, C. 1977. The contribution of paleontology to teleostean phylogeny. *Major Patterns in Vertebrate Evolution* (ed. by M.K. Hecht, P.C. Goody and B.M. Hecht), pp. 579–643. Plenum Press, New York; Patterson, C. 1981. Significance of fossils in determining evolutionary relationships. *Annual Review of Ecology and Systematics* 12: 195–223.

[56] Brundin, L. 1966. Transantarctic relationships and their significance as evidenced by midges. *Kungliga Svenska Vetenskapsakademiens Handlinger* 11 (Series 4), 1–472.

[57] Eldredge, N. and Cracraft, J. 1980. *Phylogenetic patterns and the evolutionary process: method and theory in comparative biology*. New York: Columbia University Press; Nelson, G.J. and Platnick, N.I. 1981. *Systematics and biogeography: cladistics and vicariance*. Columbia University Press, New York; Wiley, E. O. 1981. *Phylogenetics: The Theory and Practice of Phylogenetic Systematics*. New York: Wiley and Sons, Interscience.

[58] For de Candolle see Drouin, J.-M. 2001. Principles and uses of taxonomy in the works of Augustin-Pyramus de Candolle. *Studies in the History and Philosophy of Biological and Biomedical Sciences* 32: 255–275.

[59] Nelson, G. J. 1979. Cladistic analysis and synthesis: Principles and definitions, with

a historical note on Adanson's *Familles des Plantes*. *Systematic Zoology* 28: 1–21, quotation on p. 20.

[60] As does that of animalcules ("Protists"): Schlegel, M. & Hülsmann, N. 2007. Protists—A textbook example for a paraphyletic taxon. *Organisms, Diversity & Evolution* 7: 166–172.

Dear Mr. Linnaeus,

Your driving purpose was to bring order out of chaos in the living world. You gave to us a system of classification, literally *Systema Naturae* (The System of Nature) that has lasted to the present day. It can be reasonably assumed that the first words to emerge during the origin of human speech included the names of plants and animals. That advance, which may have occurred in the final emergence of *Homo sapiens* as recently as 15,000 years ago, can be regarded as the earliest forerunner of science. Accuracy and repeatability in communication about the environment were then as now necessary for survival. Getting things by their right names, as the Chinese put it, is the first step to wisdom.

During the past 2,300 years, systematics, the science of classification, evolved in Western culture through four stages. The first was the hierarchical system introduced by Aristotle. Although this first systematics of recorded history muddled the picture somewhat by strict formal criteria based on Platonic essentials, he did establish the concept of taxonomic hierarchy—in this case the *eidos* of a particular form, such as horse, dog, or lion, and the *genos*, a combination of such forms that can be regarded as part of Earth's natural order. Aristotle recognized some 520 animal species, mostly from Greece, that were consistent with his definition of *eidos*.

During the Middle Ages and into the Enlightenment, much of the research of early life scientists consisted of systematics in the Aristotelian mode, in an effort to perfect a system of classification for all known plants and animals. Microorganisms and the smaller multicellular organisms of course remained largely unknown until the invention of the microscope in the 1600s. The work of these authors, including Andrea Cesalpino, Caspar Bauhin, Joseph Pitton de Tournefort, and John Ray, culminated with the system devised during the mid-1700s by yourself, Carolus Linnaeus.

Your name is virtually synonymous with the modern era of systematics. This is because you made three decisively influential contributions. First, you presented in Leiden the *Systema Naturae* of 1735, of which I have been privileged to see your own personal copy, that formalized the hierarchical system of classification used today. In this system, a direct philosophical descendant of Aristotle's first scheme, you grouped all known organisms into three kingdoms, which were then divided successively downward into classes, orders, genera, and species. The basic unit you recognized is the species, and you aggregated the higher taxonomic categories into successively larger clusters of species according to their anatomical similarity. Although you believed in special creation, you nevertheless spent your entire career striving to define the diversity of life as a natural, comprehensible system as opposed to an arbitrary, chaotic system.

Your second major contribution was the binomial nomenclatural system. Introduced in the *Species Plantarum* in 1753 for plants and *Systema Naturae* edition 10, 1758, for animals, this system would become the standard as the starting point of biological classification. The early procedure you used was that of the very capable Joseph Pitton de Tournefort, who in 1700 characterized each genus by a single term and the species within it by a brief diagnostic description. You took the next step by simplifying the procedure with a single Latinized name for the genus coupled with a single Latinized name assigned to the species, followed by a diagnostic description. Thus we have today our own species *Homo sapiens* and our faithful companion species *Canis familiaris*. From this procedure of yours was to grow the modern rule of nomenclatural priority observed by all taxonomists: in order to ensure stability, the first binomen applied to a species with a formally published description must thereafter prevail and exclude the use of other binomens. Another stabilizing rule that arose from your procedure is the designation of types, the specimens used in the original description and as a result designated as the final reference for the species and its binomen.

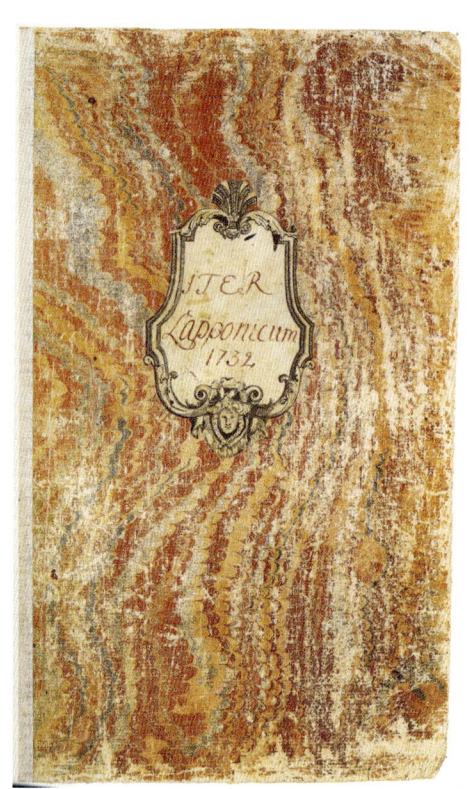

The front cover of Linnaeus' original journal of his travels in Lapland, *Iter Lapponicum*

Your binomial system facilitated your third great contribution, the initiative to find and diagnose the entirety of biodiversity, from the local Swedish biota to all those around the world. Such an effort became technically possible in your scheme because large numbers of species, including novelties, could be diagnosed and labeled efficiently. You yourself limited your field trips to Sweden, going as far as Lapland and the Baltic Island of Öland. But, ever productive in your Uppsala professorial headquarters, you inspired students, some of whom traveled widely, to collect and describe as many new species as they could find. Among your apostles, as you modestly called them, were Peter Forsskål and Karl Peter Thunberg, the first field taxonomists to visit Asia, and the pioneering insect taxonomist Johann Christian Fabricius. I'm in the process, with José Duran of Madrid, of bringing another of your apostles to the Pantheon, José Celestino Mutis, who in the late 1700s founded the Botanical Garden at Bogotá, Colombia. Most of Mutis's prodigious work was lost in shipping and war, but now Duran and I have been able to reconstruct another large piece of his newly discovered diaries and letters. The systematic exploration of the biosphere had begun, in what today can legitimately be called the Linnaean enterprise.

Where the launching of global biodiversity exploration was an 18th-century achievement, the great advance of the 19th century, the third landmark in the series of four I would like to recognize, was the introduction of evolutionary theory as the *leitmotif* of biodiversity studies. The first to promote this idea was Jean Baptiste Pierre Antoine de Monet de Lamarck, often called Chevalier de Lamarck. His *Philosophie Zoologique*, published in 1809, argued that the world's multitudinous life forms can be organized into a phylogeny, a history of ancestors and descendants.

But Lamarck's reasoning convinced few scholars of the value of phylogenetic classification or even of the fact of evolution. In fact, it ultimately turned out that his proposed mechanism of evolution was flawed. This is the theory of inheritance of acquired characters, in which changes in one generation induced by interaction with the environment tend to be passed on to the next generation.

It remained for Charles Darwin, in his masterwork, *On the Origin of Species*, 50 years later, to provide

Portrait of Charles Robert Darwin painted in 1881 by John Collier (1850–1934)

massive and compelling evidence for the ongoing process of evolution. He also put forward the correct explanation for it: natural selection, whereby spontaneous mutations create hereditary variants, which compete for survival and reproduction, resulting in the gradual replacement of some variants by others over many generations. Darwin's theory of evolution by natural selection, although at first opposed on both religious and philosophical grounds, spread in influence steadily thereafter. In time it not only succeeded but became fully transformative throughout biology. Applied to systematics, evolutionary theory cemented the concept of phylogeny and validated classification above the species level, based on phylogenetic reconstructions.

What, then, is the fourth and current advance in systematics? It is nothing less than the attempted completion of the enterprise that you began by a full mapping of Earth's biodiversity, pole to pole, bacteria to whales, at every level of biological organization from the genome to the ecosystem. It aims to yield as complete as possible a cause-and-effect explanation of the biosphere, and a correct and verifiable family tree for all the millions of species—in short, a unified biology.

This is a task which, in spite of centuries of effort already devoted to it, seems today scarcely begun. Today, almost 250 years later, we still have discovered only as few as 10 percent of the species of organisms living on Earth. Most kinds of flowering plants and birds have been discovered, but our knowledge of insects and other small invertebrates, of fungi, and bacteria and other microorganisms is shockingly incomplete. For example, about 60,000 species of molds, mushrooms, and other kinds of fungi are known to science, but the true number has been estimated to exceed 1.5 million. The number of known species of nematode roundworms, the most abundant animals on Earth, with four out of every five animals being nematode roundworms, is about 16,000 but the number could easily be in the millions. On the order of 10,000 species of bacteria are known to science, but 5,000 to 6,000 are found in a handful of fertile soil, almost all unknown to science, and some *four million* are estimated to live in a ton of soil. As far as our knowledge of them goes, they might as well be on Mars.

Each of these and millions of other species is exquisitely well adapted, and interlocked in intricate ecological webs of interaction we have scarcely begun to understand. Yet they are a large part of the foundation of the world's ecosystems. Our lives depend utterly upon this largely unknown living world.

We live, in short, on a little known planet. When dealing with the living world, we are flying mostly blind. When we try to diagnose the health of an ecosystem such as a lake or forest, for example, in order to save and stabilize it, we're in the position of a doctor trying to treat a patient knowing only 10 percent of the organs.

Now, new advances in technology, including genomics (reading the genetic codes of each species), high resolution digital photography, and internet publication, allow us to speed the exploration of the living world by as much as 10 times, and further, to organize the information and make it immediately accessible as an open source everywhere in the world.

This dream brought to reality is called the Encyclopedia of Life. Launched on May 9 of 2007, it represents the convergence of efforts by scientists principally working in museums and herbaria and other centers of the major collections of species diversity; and planners in the large libraries that contain the totality of already published information on biological diversity.

There are compelling reasons to build such an encyclopedia, with an indefinitely expansible page for each species containing everything known about it. New phenomena, and new connections among phenomena will come to light. Only with such encyclopedic knowledge can biology as a whole fully mature as a science and acquire predictive power species by species, and ecosystem by ecosystem.

The Encyclopedia of Life will serve human welfare in immediate practical ways. The discovery of wild plant species adaptable for agriculture, new genes for enhancement of crop productivity, and new classes of pharmaceuticals, all will be accelerated. The outbreak of pathogens and harmful plant and animal invasives will be better anticipated. Never again, with knowledge of adequate extent, need we overlook so many golden opportunities in the living world around us, or be so often surprised by the sudden appearance of destructive aliens that spring from that world.

Those of us who planned the Encyclopedia of Life are grateful for the support now promised by the MacArthur and Sloan Foundations for this initiative, and by the favorable attention we hope the wider public will give it. We are also grateful to the memory of you who led the way in the systematic exploration of life on Earth.

Your humble apostle,

Edward O. Wilson

316 *Letters* to *Linnaeus*

Earth from space—the "little known" planet

Linnaean works cited by the correspondents

Artedi, P. 1738. *Ichthyologia*. Edited by C. Linnaeus. Leiden.
Bru, J.B. 1784–1786. *Colección de Láminas que representan los Animales y Monstruos del Real Gabinete de Historia Natural*. Madrid.
Catesby, M. 1730–1747. *The Natural History of Carolina, Florida and the Bahama Islands*. London.
Clerck, C. 1757. *Svenska Spindlar*. Stockholm.
Fabricius, J.C. 1775. *Systema Entomologica*. Flensburg & Leipzig.
Fabricius, J.C. 1778. *Philosophia Entomologica*. Hamburg & Kiel.
Fabricius, J.C. 1805. *Systema Antliatorum*. Brunswick.
Giseke, P. 1792. *Praelectiones in Ordines Naturales Plantarum*. Hamburg.
Linnaeus, C. 1735. *Systemae Naturae*. Ed. 1. Leiden.
Linnaeus, C. 1736. *Fundamenta Botanica*. Amsterdam.
Linnaeus, C. 1736. *Musa Cliffortiana*. Leiden.
Linnaeus, C. 1737. *Critica Botanica*. Leiden.
Linnaeus, C. 1737. *Flora Lapponica*. Amsterdam.
Linnaeus, C. 1737. *Genera Plantarum*. Leiden.
Linnaeus, C. 1738 [1737]. *Hortus Cliffortianus*. Amsterdam.
Linnaeus, C. 1738. *Fragmenta Methodis Naturalis* in *Classes plantarum*. Leiden.
Linnaeus, C. 1745. *Hortus Upsaliensis*. Stockholm.
Linnaeus, C. 1751. *Philosophia Botanica*. Stockholm.
Linnaeus, C. 1753. *Species Plantarum*. 2 vols. Stockholm.
Linnaeus, C. 1758–1759. *Systemae Naturae*. Ed. 10. 2 vols. Stockholm.
Linnaeus, C. 1766–1767. *Systemae Naturae*. Ed. 12. 2 vols. Stockholm.
Linnaeus, C. 1767. *Mantissa Plantarum*. Stockholm.
Linnaeus, C. 1771. *Mantissa Plantarum Altera*. Stockholm.
Linnaeus, C. 1788–1793. *Systemae Naturae*. Ed. 13. 3 vols. Edited by J.F. Gmelin. Leipzig.
Loefling, P. 1758. *Iter Hispanicum*. Stockholm.
Olivi, G. 1792. *Zoologia Adriatica*. Bassano.

Notes:

[1] For complete lists and more details on Linnaean bibliography with locations of copies Linnaeus' published works see the Linnaeus Link project at http://www.linnean.org/; also see Blunt, W. 2001. *The Compleat Naturalist: a life of Linnaeus*. Frances Lincoln, London, and Jarvis, C. 2007. *Order out of Chaos: Linnaean plant names and their types*. Linnean Society of London and the Natural History Museum, London.

[2] To view the Linnaean Correspondence in its entirety, and the Linnaean Collections of insects, fish, shells and plants held at the Linnean Society of London please visit the Society's website http://www.linnean.org, click on "Linnaean Collections Online" and choose from the collections menu.

Picture credits

Every effort has been made to source the correct copyright for each image. The publishers will be happy to make good any errors or omissions brought to their attention in future editions.

© Linnean Society of London: ii, viii, 4, 11, 14, 15, 18, 20–23, 27, 28–30, 36, 40, 48, 52, 57, 58, 63, 66, 67, 76, 80, 84, 85, 86, 90, 92–94, 101–104, 108, 110, 113, 116, 117, 120, 121, 128–131, 136 (both images), 138, 141 (both images), 145, 146, 149, 152, 155 (both images), 158, 162, 164, 167–169, 173 (both images), 182, 188–190, 193, 194, 201, 204, 206, 211, 212, 214, 215, 220 (all images), 221, 223, 224, 226, 228, 232, 236 (all images), 240, 241, 243, 246, 248–250, 254, 256, 258, 232, 236 (all images), 240, 241, 243, 246, 248–250, 254, 256, 258, 259 (both images), 262, 263, 264 (all images), 266, 267, 279, 282, 287–290, 292, 300, 303, 312, 313

© Sandra Knapp: 32, 37, 41, 43, 59, 81, 99, 111, 114, 125, 147, 170, 237, 245

© Linnean Society of London/Leonie Berwick: 49, 71, 82, 106, 185, 207, 223, 252, 274

© Pat Morris: 1, 77, 123, 159, 178, 200

© Andrew Polaszek: xii

© ⓘ Design attributed to Donat Agosti: 3

© John Alcock: 7, 9

© Jack Longino, Evergreen State College: 12

© Goran Henriksson, Uppsala Astronomical Observatory: 33

© David Cutler: 45–47

© Sam Stubbs: 60

© NASA (www.nasaimages.org): 87, 316

© Dr. Diana Lipscomb, George Washington University: 126, 291

© Ellinor Michel: 135

© Brooklyn Botanic Garden: 153

© American Museum of Natural History (AMNH), by permission of Norm Platnick: 177

© Dr. G. B. Edwards: 183

© Leonie Berwick: 186

© Tamara L. Clark (www.tamaraclark.com): 195

© Uppsala Universitet, Sweden, by permission of Gunnar Tibell: 271

© Museo Nacional de Ciencias Naturales, Madrid: 276

© R. I. Vane-Wright: 284 (all images)

© Charles Kazilek, Arizona State University: 294

© Natural History Museum, London: 297, 302

Logos

Royal Botanic Garden Edinburgh logo used by permission of Stephen Blackmore: 25

Biodiversity Institute, University of Kansas logo used by permission of Leonard Krishtalka: 31

Encyclopedia of Life logo and webpage used by permission of James L. Edwards: 53, 55

California Academy of Sciences logo used by permission of David P. Mindell: 139

The New York Botanical Garden logo used by permission of Dennis Wm. Stevenson: 251

Index

Italic pagination refers to illustration captions.

Abramites hypselonotus 232
Acalles 12
Acherontia atropos 214, 215
Achillea alpina 140, *141*
   A. ptarmica 140, *141*
Acipenser 157
Acrodonta 179
Adanson, Michel 231
affinity 17, 299–301
Agama 178
Agamidae 178–81
Agardh, Carl Adolph 298–9
Agassiz, Jean Louis Rodolphe 171, 191
Aglais urticae 283
Aira carophyllea 154
   A. praecox 154
algae 298–9
All Species Foundation 1
Alosa 157
   A. sapidissima 163
Alstrin, Laur 232
Altingiaceae *47*
Amanita muscaria 246
Amazon 285
   deforestation *114*
American Museum of Natural History 171
Ammodytes 157
Amoenitates Academicae 108
Amphibia 200
anagenesis 150
Anales de Historia Natural 275, 277
analogy 299–300
Ananas comosus 226
Anarhicus 157
Anchonus 12
Anemone canadensis 153
Annelida 77
Anthoxanthum odoratum 154
Aphorismi Botanici 299
apostles *see* students of Linnaeus
Appias 284
Arabidopsis (Arabis) thaliana 155
arachnology ix, 65–6, 171–4, *177*, 183–4
archaeology 46
Arduino, Pietro 147
Argentina 9
Argentina (genus) 157
Ariidae 182
Ariopsis felis 182
Aristotle 207–8, 209, 240, 311
Artedi, Peter 157, 159, 160, 161, 163, 191
artificial classifications 160–1, 229–30, 231, 240, 297, 301
Aster dumosus 154
   A. novi-belgii 154
Asteraceae *141*, 240
Astragalus canadensis 153
Atherina 157
Atriplex patula 221
Atta 1
Attacus atlas 215
Australia 95, 160, 166, 249

Bacillaria 298
Balistes 157
Banks, Joseph 93, *249*
Bates, Henry Walter 281, 284, 285
Bauhin, Caspar 311
Berlese, Antonio 11; Berlese funnels 11–12, *12*
Berlin, Anders 21–4
binomial nomenclature 11, 16, 53–4, 85, 94, 123–4, 169, 205, 235, 247, 259, 295, 312
   authorship abbreviations 53, 154–5, 216, 219–20
   enduring value of ix, xi, xii, 5, 10, 61, 48–9, 74, 151, 154, 241, 260, 293
   and fossils 59–62, 123
   problems with 42–3, 67–9, 125, 134, 139–40, 163
   species names
      changing names 43
      and geographical distribution 163, 165, 209, 241
      new names 260, 273
      synonyms 216
   universality of 225, 260
binomial nomenclature, challenges to 36–9, 61, 117–18, 154, 158–9
see also cladistics; PhyloCode *and* quinarianism
biodiversity
   crisis xi, 1, 2, 112, 125, 197, 233, 295
   magnitude of x, 12, 16, 32, 41, 53, 74, 77, 95, 112, 125, 149, 151, 185, 195–6, 205, 219, 227, 229, 246, 283, 293, 305
Biodiversity Heritage Library (BHL) 55
biogeography 165
Biosystematic Database of World Diptera 264
Blattodea 199, 202
Blennius 157
Blumenbach, Johann 105
Blunt, Wilfrid 263
Boisbaudran, Paul-Emile Lecoq de 89
Bonnet, Charles 209
botany 25–7, 41–7, 57, 85, 95, 111–15, 129–32, 140–1, 153–5, 161, 169–70, 189, 225–6, 229, 239–42, 275, 300–1
Bradley, John 217
Brazil 166
Brisson, Mathurin Jacques 191
Bromelia ananas 226
Bromus tectorum 154
Brongniart, Aldophe Théodore 61
Brookes, Richard 196
Bru, Juan Bautista *276*
Brucharachne 172
Buddleja 282
Buffon, Comte de *see* Leclerc, Georges-Louis
Burlini, Milo 147
Burns, John 171
Butterflies and Moths of the Wayside and Woodland 282

Callionymus 157
Camellia sinensis 237
Canada 104, 225
Candolle, Augustin Pyramus de 160, 230, 241, 297–8, 300, 301, 304
Canis familiaris 312

Caribbean 132
  Bahamas 29
  Jamaica 58
Carpenter, Jim 176
Carr, Archie 159, 183
Catalogue of Life 1, 170
Cavanilles, José Antonio 276
Celsius, Anders *33*, 253
  Celsius scale 31–2, 155, 253
Centriscus 157
Centropristis philadelphica 18
Cesalpino, Andrea 311
Chaetodon 157
Chainey, John 284
Chamaeleonidae 178–81
Champion, George C. 11
Chaos 298
Characinidae 163
characters, generic 17, 173–4, 176, 209, 228
Cheilopogon exsiliens 164
Chelsea Physic Garden, London, UK 57
Chenopodiaceae *221*
Chile 166
China *37, 237*
Chironoma plumosa 259, 260
Christian of Lyons 31
Chromis abyssus 195
Ciliata *292*
cladistics xi, 36, 38, 54, 63–73, 100, 150, 172, 175–9, 180–1, 199–203, 208, 231, 293, 304–5
cladogram xi, 177
Cladotanytarsus marki 260
classification
    pre-Linnaean 311, 312
    post-Linnaean 297–310, 313–14
Cleonis piger 63
Cleora 216
Clerck, Carl Alexander ix, *173*, 221, 281, 285
Clifford, George 57, 225, 255
climate change 1, 13, 25, 26, 97, 112
Clitoria mariana 153
Clupea harengus 157
Clupeidae *158*
Cobitis 157
Coelosis biloba 246
Colaptes auratus 232
Coleophora vestianella 222
Colliculoamphora reedii 297
Collins, Francis 197
Colombia 275, 313
common descent 96, 139, 142, 144, 176, 177, 186, 192, 202, 302
comparative biology 16
computers, use of 1, 2, 54, 78, 95, 100, 185, 205, 229, 247; *see also* internet
conchology *136*, 190
congresses
  Congreso Nacional de Botánica, XII (Puerto Maldonado, Peru; 2008) 111, 114
  International Botanical Congress, XVI (St Louis, USA; 1999) 121
  International Congress of Zoology, XX (Paris, France; 2008) 185, 187
consensus taxonomies 74, 78
conservation 2, 97, 169
Conspectus Criticus Diatomacearum 299
Convention on Biological Diversity 114, 169
convergent evolution 96
Copepoda *303*
Copernicus, Nicolaus 88
copyright 1, 2

Coregonus 157
Cornus florida 153
Corvus corax 206
  C. monedula 66, *67*
Coryphaena 157
Cottus 157
Crabronidae 7
creationism xi,192–3
Crick, Francis 197
Critica Botanica 119, *120*, 255
Croizat, Léon 300
Crosskey, Roger 219
Cruciferae 300
Cryptogamia 298
Curculio sulcirostris 63
Curculionidae 11–12
Cuvier, Georges 16, 61, 191, 196
Cyanocitta cristata 153
Cyclocephala amazona 246
Cyclopterus 157
Cydia pomonella 220
Cynodon dactylon 154
Cyprinidae *194*
Cyprinus 157

Dahlgren, Johan 111
Dahlgren, Rolf M.T. 241
Dalibarda repens 153
Danaus plexippus 153
Darwin, Charles 31, *36*, 54, 77, 96, 124, *149*, 150, 192, *193*, 207, 208, 227, 236, 276, 282, 293, 295, 304, *313*
  on Linnaeus 302
deforestation 97, *114*
Denmark 239
Dicrotendipes thanatogratis 260
digital imagery 194–5, 222, 315
Diodon 157
Dioptrophorus 12
Diospyros kaki 226
Diplodocus carnegiei 123
Dissertationes 189, *252*
DNA 54, 161, 228; *see also* molecular sequencing
Du Petit Thouars, Louis-Marie Aubert 241
Dubois, François-Noël-Alexandre 230
Duran, José 313

Echeneis 157
ecology 268, 302, 314
Ehrenberg, Christian Gottfried 298, 306
Ekström, Daniel 31
Ellis, Brian 209
Ellis, John 145
Elvius, Pehr 31
Encrasicholus 157–8
Encyclopedia of Life (EOL) 1, 54–6, 170, 264, 315
endemism 95, 165
entomology *1*, 5–10, 11–12, 63, 69–71, 147, 199–200, 202, 259–61, 263–4, 273
Entomostracites 61
Epler, John 260
Equatorial Guinea 24
Erbenochile 60
Erythrina crista-galli 58
Eschmeyer, Bill 183
Esox 157
essentialism 160, 176, 207, 208–9, 210, 243
Eurhoptus 12
evolution 139, 142, 186, 202, 267
  principle of 54, 96, 192, 207, 227, 276
  process of 124–5, 314

Exocoetidae *164*
Exocoetus 157
extinction of species 13, 32, 43, 56, 61–2, 125, 155, 196–7, 206, 233, 281–2, 284, 293, 295; *see also* fossils

Fabaceae *52*
Fabricius, Johann Christian 217, 219, 239, 263, 281, 286, 313
Faraday, Michael 99
Felis volantis 290
Felsenstein, Joe 174, 297
Fiji 216
The Fishes of Alachua County, Florida 159, 183
Fistularia 157
Flora Lapponica 255, *256*
Flore Française (de Candolle) 230, 297, 300–1
Flore Françoise (Lamarck) 230, 301
Forsskål, Peter 16, 313
fossils 59–62, 123, 227, 304
Foucault, Michel 107
The Foundations of Systematics and Biogeography 300
Fragmenta Methodi naturalis 297, 305
France 185, 187, 223–4
Franklin, Rosalind 197
Franklinia alatamaha 43
Fraxinus excelsior 46
Freeman, Paul 283, 286
Fries, Elias Magnus 299
Fundamenta Botanica 119, *120*, 255
funding for taxonomic work 3, 17, 32, 252, 294
Furia 298
Furnarius rufus 66

Gadus 157
Gallica 160
Gasterosteus 157
Gaultheria procumbens 153
Gauthier, Jacques 176, 180
GenBank 2
Genera Plantarum 229, 255
genetic variation 109
genome research 2, 126, 142–3, 197
Gentiana autumnalis 153
 G. sino-ornata 37
geographical distribution
 and species names 163, 165, 209, 241
Geological Society of London 61
geology 61, 267, 281
Gertsch, Willis 172
Gesner, Conrad *82*
Gilbert, Charles H. 191
Giseke, Paul Dietrich 241, *242*
Global Biodiversity Information Facility (GBIF) 2, 170, 264
Global Strategy for Plant Conservation (GSPC) 95, 114
Gobias 157
Gobio gobio 194
Goethe, Johann Wolfgang von ix, 145
Gosline, William A. 191
Gould, Steve 171
Grey, Thomas de 220
The Growth of Biological Thought 207
Guatemala 12
Gyllenhal, Leonard 221
Gymnotus 157
Gymnura altavela 15
Gyrinidae *294*

habitat destruction 233; *see also* deforestation
Haeckel, Ernst *201*, *292*, *302*, *303*

Haller, Albrecht von 227
Hampson, George 217
Harvard University, USA 171
Hayes, Alan 217
Helonias bullata 153
Hemipepsis ustulata 5, 10
Hennig, Willi 16, 35, 36, 78, 100, 178, 208, 284, 286, 293, 295
herpetology 176, 199
Herrich-Schäffer, Gottlieb August Wilhelm 219–20
hierarchy, Linnaean xi, 37, 68, 174–5, 247
 see also ranks, taxonomic
Histoire naturelle générale et particulière 148
Historia Animalium 82
Holman, Eric W. 278
Holorusia 286
Homo sapiens 311, 312
 taxonomy of subspecies 34, 105–9
homology 302; *see also* affinity
Hooker, Joseph 299
Hoppius, C.E. *108*
Hortus Cliffortianus 57, *58*, 255
humans in Linnaean hierarchy 105, 267
Humboldt, Alexander von 275–6
Hummel, Arvid David 221
Hunter, John 275
hybridization 96, 140, 210, 286
Hydra 298

Ichthyologia 15, 157, 159
ichthyology 15–18, 86, 157–9, 163, 191–2
Icones Insectorum Rariorum 285
information science 99
 post-Linnaean advances in 229, 236–7
Insects of Samoa 217
Institutions of Entomology 285
International Association for Plant Taxonomy (IAPT) 227
International Code of Area Nomenclature (ICAN) 165
International Code of Zoological Nomenclature 120, 134, 218
International Commission on Zoological Nomenclature (ICZN) 119, 121, 133, 134–6, 227
International Society of Phylogenetic Nomenclature 208
internet 194, 230–1
 Biosystematic Database of World Diptera 264
 Catalogue of Life 1, 170
 Encyclopedia of Life 1, 54–6, 170, 264, 315
 Gallica 160
 GenBank 2
 Google 1, 2, 33, 160
 Linné online 268–71
 World Spider Catalog 183
 ZooBank 137, 195
introduced species xi, 232–3
Isoptera 199–200, 202
Italy 59, *147*, 148
Iter Hispanicum 276
Iter Lapponicum 117, *155*, *312*

Jacquin, Nikolaus von 263
Japan 268, 272, 284
Jones, William 281, 283, 285
Jordan, David Starr 191
Juniperus virginiana 153
Jussieu, Antoine-Laurent de 227, 230, 241
Jussieu, Bernard de 111, 191, 227
Jussieu, Joseph de 111

Kalm, Pehr 103–104, 111, 153
Kalmia angustifolia 153
  K. latifolia 153
Kertész, Kalman 263
Klingenstierna, Samuel 270, *271*
Kudo, Richard Roksabro 291
Kullander, Sven 119
Kunstformen der Natur 201, *292*, *303*
Kützing, Friedrich Traugott 298

La Mettrie, Julien Offray de 150
La Ville-sur-Illon, Bernard Germain Étienne 191
Labrus 157
Lacandonia schismatica 228
Lacépède, Comte de *see* La Ville-sur-Illon, Bernard Germain Étienne
'Lacertilia' *201*
Lamarck, Jean-Baptiste 191, 192, 230, 301, 313
Lamiaceae *170*
Lamium album 170
languages used in science 119, 124, 160, 218, 225, 226–7, 235, 268
Laplace, Pierre-Simon, Marquis de 87
Lawrence, Peter 277
Leclerc, Georges-Louis 19, 148, 191, 227
Leeuwenhoek, Antonie van 298
Leiolepis 178
Lemurus caudatus 290
lepidopterology 213–222, 235, *236*, 237, 281–6
Levi, Herbert 171
L'Héritier de Brutelle, Charles Louis 241
Linaria 228
Linnaea borealis 153
Linnaeus, Carl
  anniversary celebrations 271–2
  clothing typical of *233*, 257
  drawings by *117*, *211*
  house at Hammarby, Sweden 233, *267*, 270, *271*
  medical work 50, 252, 267
  memorials to 25
  personality 257
  pet raccoon *232*
  religious beliefs 16, 105, 140, 141, 142, 210, 240, 293, 312
  research on food sources 1, 226, 237
  specimens collected for 24, 26, 29, 58, 85, 111, *132*, *167*, 235, 245, 246, 249, 251, 273
  statues *207*, 268
  teaching 251–3, 257
  travels *211*, 239, 270, 312
  websites about 268–71
Linnaeus, Carl, filius 86, *131*, 189, 190, 246, 275
Linné, Carl von *see* Linnaeus, Carl
Linné online 268–71
Linnean Society of London
  collections 47, 190, 220–1, 233–4, 237, 268
    fishes *194*
    insects *11*, *71*, *220*, *236*, *264*, *274*
    plants *27*, *52*, *76*, *113*, *141*, *145*, *155*, *221* 228, *240*
    shells *136*
  foundation of 93, *94*
  library *49*, *185*, *220*, *252*
Liquidambar styraciflua 47
Lobelia dortmanna 153
Loefling, Pehr 111, 275, 276, *279*
López González, Ginés A. 280
Lophius 157
Loricaria 157
Löther, Rolf 210
Lumbricus terrestris 77

MacArthur Foundation 315
MacLeay, William Sharp 299, *300*
Macrolepidoptera of the World 215
Maillet, Benoît de 139
Maniola jurtina 283
Mantissa Plantarum 164
Maryella reducta 260
Massonia latifolia 131
Maupertuis, Pierre-Louis Moreau de 87, 88
Mayr, Ernst 35, 171, 207, 278
Mello-Leitão, Candido Firmino de 172
Mendel, Gregor 197
Mendeleev, Dmitri 88, 89, 90
Mephitis mephitis 232
Mexico 12, 166, 225, 228, *245*, 246–7
Mickevich, Mary 181
microscopic organisms ix, 53, 74, 79, 125, 126, 142, 291, 297–9
microscopy 15 7, 126, 297
Milium effusum 153
Miller, Philip 57, 58
mimicry 281, 285
Mitchella repens 153
molecular sequencing 79, 95–6, 126, 127, 231, 260, 264
Monarda clinopodia 153
monogenesis 105
Montpellier Botanic Garden, France 85
Mormyrus 157
Morus 232
Mugil 157
Mullus 157
Muraena 157
Murray Fundamenta Testaceologia 190
Mus volans 290
Musa Cliffortiana 226, 255
Musa paradisiaca 225, *226*
Musca vespiformis 264
mutation 96
Mutis, José Celestino 275, 313
mycology 117, 227, *246*
Myers, George 191
Myrmecophaga tridactyla 276
mythological creatures 51, *82*, 83, 107, 235, *236*, 239

Naef, Adolf 304
Nannarrup hoffmani 155
natural classifications 81–3, 160, 161, 229–30, 231, 240, 241, 297, 299–302
Natural History Museum, London, UK 11, 101, 127, 217, 220, 283
natural selection 96, 192, 227, 314
Nautilus pompilius 136
The Nearctic Species of Tendipedini [Diptera, Tendipedidae (=Chironomidae)] 259
Nelson, Gareth 304
Nelumbo nucifera 125
Nessaea obrinus 285
Netherlands 58, 255, 268
'New Systematics' 171, 207
Newton, Isaac 87–8, *90*
Nicrophorus americanus 55
Nielsen, Ebbe 220
numerical taxonomy 231
Nymphaea alba 228
Nymphalis io 283

Observationes in Regnum Animale 157
Oldroyd, Harold 283, 286
Oleaceae *46*
Olivi, Giuseppe 147, 148

On the Nature of Limbs 302
On the Origin of Species by Means of Natural Selection
  36, *193*, 302, 313–14
ontogeny 302
Ophidion 157
Orbigny, Alcide d' 306
Order out of Chaos 101
Orectochilus orbisonorum 294
ornithology 66–7, 191, *206*, 230
Orthosia 215
Oryza sativa 145
Osiander, Andreas 88
Osmerus 157
Ostracion 157
Owen, Richard 300, 302

Pachybrachis 147
palaeontology 59–62, 298, 304
Panicum dactylon 154
Papilio 284
  P. ancaeus 281
  P. atalanta 236
  P. obrinus 281, 285
Paramerina smithae 260
Passer domesticus 66, 177, 232
Pegasus 157
Peloria 228
Peprilus alepidotus 167
Perca 18, 157
Peru *99*, 111–15
Petromyzon 157
Phalaena vestianella 221
Phaseolus vulgaris 52
pheneticism xi, 38, 293
Philosophia Botanica 112, *223*, *224*, 227, 241
Philosophia Entomologia 239
Philosophie Zoologique 313
Philosophy of the Inductive Sciences 301
The Philosophy of Nature 209
Phlox pilosa 153
PhyloCode 18, 150, 175–82, 199–200, 202, 208, 227, 231, 291
phylogenetic systematics *see* cladistics
phylogeny 46, 302, 313
Pieris brassicae 283
  P. rapae 283
Pitta sordida 66
Placus 291
Plagiopyla 126
plant micromorphology 45
Plantago 232
Platnick, Norm 304
Pleijel, Fred 172
Pleuronectes 157
Plot, Robert 196
Poa annua 154
  P. bulbosa 154
Pogonomyrmex pima 5, 10
Polynemus 157
Pomacentridae *195*
Pomatomus saltatrix 163
Popper, Karl 180
population, human 1, 13, 25, 56, 61, 185–6, 197, 281
predictions in classification 181, 184, 315
Procyon lotor 232
Prorodontidae *291*
Protapion trifolii 11
Protea lepidocarpodendron 76
Protozoology 291
Pulawski, Wojceich 7, 8, 10
Pulle, August Adriaan 241

Queiroz, Kevin de 176, 180
Quercus alba 153
  Q. phellos 153
quinarianism xi, 35, 293, *300*

radiometric dating 144
Raja 157
Randall, John E. 191
rankless nomenclature 158–9
ranks, taxonomic 199–200
Rauwolfia 58
Ray, John 311
reductionism 83
Reptilia 200
Rhea americana 163
Rhus copallina 153
Ribes cynosbati 153
Riley, Norman 283, 286
Rothman, Johan 242
Royal Botanic Garden Edinburgh, UK 25–6
Royal Botanic Gardens, Kew, UK 93, 98, 132
Rydbeck, Olof 97

Saint-Hilaire, Augustin 299
Saint-Hilaire, Étienne Geoffroy 191
Salmo 157
Salticidae 183, 184
Sandoricum koetjape 226
Sapotaceae *30*
Sattler, Klaus 219
Saturniidae *215*
Scarabeus sacer 121
Scheuchzeria palustris 153
Schiller, Friedrich 81
Sciaena 157
scientific discoveries, post-Linnaean 87–91, 99
scientific information
  accessibility 1, 2, 54–6, 77, 78, 160, 183, 194–5
scientific methods and technology, post-Linnaean 222, 228, 295, 315; *see also* computers; digital imagery; microscopy; molecular sequencing,; radiometric dating *and* staining techniques
scientific papers
  evaluation 277, 279
Sclater, Philip Lutley 165, 166
Scomber 157
Scorpaena 157
Seba, Albertus 15, 192, 239, *290*
Seitz, Adalbert 215, 216
Serranidae 18
sexes, symbols used for *155*
sexual system of plant classification 25, 57, 58, 140, 150, 154, 161, 227, 229–30, 239–41, 297
Seymer, Henry 236
Shaffer, Michael 217
Sibbald, Robert 26
Sibbaldia procumbens 26, *27*
Sideroxylon foetidissimum 30
Siegesbeck, J.G. 240
Siegesbeckia occidentalis 240
Silphidae *55*
Silurus 157, 159
Smeathman, Henry 24
Smith, James Edward 93, *94*, 190
Smith, Ken 283, 286
Smithsonian Institution, USA 166, 199, 203, 264
Solanaceae 113
Solander, Daniel 16, 190
Solanum montanum 112
  S. peruvianum 112, *113*

Solidago canadensis 154
   S. sempervirens 154
South, Richard 214
South Africa 95, 132
Spain 268, 275–7, *279*
Sparrman, Anders 132
Sparus 157
Species Plantarum ix, 41, 43, *85*, *114*, 139, 149, 150, 154, *169*, 205, 255, 312
species, number of *see* biodiversity: magnitude of
Sphingidae *214*, *282*
Sphinx ligustri 282, 283
Spongia 298
Spratti 157
Sprattus sprattus 158
Sprengel, Kurt Polycarp Joachim 301
Squalus 157
staining techniques 17, 46
Stainton, Henry 217
statistical frequency methodology 278–9
Stephens, James 217
Strindberg, August ix
Stromateus 157
Strömer, Mårten 31
Struthio camelus 177
students of Linnaeus 16, 21–4, 103–4, *108*, 111, 132, 153, 190, 217, 219, 235, 239, 249–51, *252*, 263, 268, 275, 276, *279*, 281, 286, 294, 313
students, modern-day *111*, 112, 160
Sturnus vulgaris 232
Svenska Spindlar xii, *173*
Swainson, William 31
Sweden 187
Swofford, David 100
Syngnathus 157
Syrphidae *264*
Systema Algarum 298, 299
Systema Antliatorum 263
Systema Entomologiae 239
Systema Mycologicum 299
Systema Naturae 34, 77, 81, 89, 119, 140, 149, 205, 207, 239, 246, 255, 311
   1st ed. 157, 159
   10th ed. *viii*, ix, *15*, 53, 70, 105, 123, 154, 157, 159, *173*, *185*, 192, 193, *259*, 263, 312
   12th ed. 16, 163, 229, 298
   13th ed. 148
Systematic Biology 172
Systematic and Evolutionary Biogeographical Association (SEBA) 165
systematics, phylogenetic *see* cladistics
systematics, trends in 18
Systematics and Biogeography 304

Tachytes 10
   T. amazonus 9
   T. distinctus 9
   T. ermineus 7, 8, 9
   T. intermedius 9
   T. sculleni 8, 9
   T. spatulatus 8, 9
   T. tricinctus 9
Tamias striatus 30, 153
Tams, Timothy 217
Taraxacum 232
Taxus brevifolia 80
Temnostoma vespiforme 264
Tenebrionidae 273, *274*
Termitidae 200
Tetraodon 157

Thalictrum flavus 140
   T. lucidum 140
Thamnophiluis doliatus 66
Theaceae *43*, 237
Theobroma 226
Theognete 11
   T. laevis 12
Théorie élémentaire de la Botanique 297, 298, 300, 301, 304
Thunberg, Carl Peter 132, 268, 313
Tineola bisselliella 221
Tipula plumosa 259
Tipulidae *259*, 286
Tortricidae *220*
Tournefort, Joseph Pitton de 311, 312
Trachinus 157
Tree of Life 54, 144
Trichiurus 157
Trifolium reflexum 153
Trigla 157
Turdus migratorius 153, 232
Turton, William 217
Tylodinus 12
type specimens 63, 134, *135*, 237, *264*, 312
Tyrannosaurus rex xii, *123*
Tyrannus tyrannus 66

United States of America 5–10, 29, *41*, 153, *167*
Uppsala University, Sweden 221, 233, 267, 271, 272
Uranoscopus 157
Uvularia sessifolia 153

Vahl, Martin 239
Valeriana 241
   V. rubra 229
Vandelli, Domingos 221
Vanessa atalanta 236, 283, *284*, 285
   V. indica 284
variation, interpretation of 17
Venezuela 275
Venter, Craig 79, 197
Verbesina occidentalis 240
Vermes 77, 298
Vietnam *186*
Vitis vinifera 45
Volta, Alessandro 99
Volvox 298
Vorticella 298
Vroeg, Adriaan *290*

Wallace, Alfred Russel 165, 166, 192
Watson, Allan 217
Watson, James 197
White, Gilbert 236
Winsor, Mary P. 160, 210, 278
World Spider Catalog 183

Xiphias 157

Yama tahetiensis 260
Yata, Osamu 284
Yeats, Thomas Pattinson 281, 285

Zamia furfuracea 245, 246
Zea mays 155
Zenaida macroura 232
Zeus 157
ZooBank 137, 195
Zoologia Adriatica 147